T0186519

PROSPECTS FOR PASTORALISM IN KAZAKSTAN AND TURKMENISTAN

Dismantling the Soviet Union had rarely-examined effects on remote rural families. Nomadic pastoralists had been collectivised into state farms to build industrialised livestock production. In a 'second Revolution', independence in 1991 brought new policies removing most state support and control. Pastoral families now choose how to manage rangeland and livestock, but with options restricted by deteriorating economies. Few pastoralists may survive this restructuring, yet they possess skills of irreplaceable value for rangeland and livestock management.

The book documents these impacts in Kazakstan and Turkmenistan, through multidisciplinary field studies during 1998–2000. Topics covered: agrarian reform; vegetation dynamics; livestock nutrition, productivity, grazing patterns, marketing and income; land degradation; institutions for managing pasture and animals. The fourteen authors, of whom five are Central Asian, include social anthropologists, animal scientists, economists, pasture agronomists and climate ecologists.

State farm dissolution has encouraged a variety of new institutions for managing livestock, land and markets with emerging socio-economic differentiation due to unequal access to resources and markets. Destocking and reduced livestock mobility has altered the locus of overgrazing, which was widespread in the Soviet period, with some pastures regenerating.

Contrasting prospects for pastoralism are drawn for each country. Swift decollectivisation in Kazakstan resulted in massive destocking and impoverishment. The future for small-scale pastoralists in Kazakstan is not promising. The Turkmen government has pursued a gradualist policy that may be more beneficial for pastoralists. The book offers suggestions for strengthening the pastoral economies.

The book will appeal to those interested in how pastoralists have coped with radical change, to researchers on range and livestock in semi-arid areas, and to policy analysts of transition in the former Soviet Union.

Carol Kerven is a Research Associate at Macaulay Institute, Aberdeen.

CENTRAL ASIA RESEARCH FORUM
Series Editor: Shirin Akiner
School of Oriental and African Studies, University of London

Other titles in the series:

PROSPECTS FOR PASTORALISM IN KAZAKSTAN AND TURKMENISTAN

From state farms to private flocks

Edited by Carol Kerven

RoutledgeCurzon
Taylor & Francis Group

LONDON AND NEW YORK

This book is dedicated to the shepherds of Central Asia,
and to those who work on their behalf

First published 2003
by RoutledgeCurzon, an imprint of Taylor & Francis
2 Park Square, Milton Park, Abingdon, Oxon OX14 4RN

Simultaneously published in the USA and Canada
by RoutledgeCurzon
711 Third Avenue, New York, NY 10017

RoutledgeCurzon is an imprint of the Taylor & Francis Group

Typeset in Baskerville by Keystroke, Jacaranda Lodge, Wolverhampton

British Library Cataloguing in Publication Data
A catalogue record for this book is available from the British Library

Library of Congress Cataloging in Publication Data
A catalog record for this book has been requested

ISBN 0–700–71699–8

CONTENTS

NOTES ON CONTRIBUTORS

Ilya I. Alimaev Scientific Director, Kazak Institute of Pasture and Fodder. Pasture agronomist with twenty-five years of research experience in management and improvement of natural pastures in Kazakstan; 65 publications, including some in English.

Roy H. Behnke Jr. Pastoral Development Officer, Macaulay Land Use Research Institute, Aberdeen. Social anthropologist specialising in evolution of land tenure systems and social institutions for controlling landed property among pastoralists. Research in Kazakstan and Turkmenistan 1998–9; 24 academic publications (including several books).

Jim Ellis Died in 2002. Senior Scientist, Natural Resources Ecology Laboratory, University of Colorado. Climate ecologist specialising in the integration of remote-sensed vegetation index data with simulation models in GIS context, to assess the climate-driven dynamics of dry rangeland ecosystems and the patterns and impacts of grazing in/on those systems.

Ovlyakuli Hodjakov Died 2001. Director until 1998 of Institute of Pasture, Livestock and Vetinary Development, Turkmenistan. Research specialisation on breeding astrakhan (karakul) sheep adapted to desert environment of Turkmenistan, under conditions of migratory stock-husbandry; 35 publications.

Carol Kerven Research Associate, Overseas Development Institute, London. Social anthropologist and Central Asian project team leader, specialising in extensive livestock production systems in semi-arid areas, on pastoral livestock marketing and household decision-making on grazing management. Research on livestock systems in Central Asia from 1995 to present; 22 publications; editor of several books and journals.

Khodja Khanchaev Senior Scientist, Institute of Animal Husbandry, Turkmenistan. Research on process of land degradation around settlements and water points in Turkmenistan's rangelands, measuring changes in vegetation communities related to grazing pressure.

Re-Yang Lee Professor, National Taiwan University, Taipei.

Christopher Lunch Research Assistant, Overseas Development Institute, London. Social anthropologist, fieldwork in Kazakstan and Turkmenistan 1998–9. Research on institutional reorganisation of livestock farming after privatisation.

Nurlan I. Malmakov Senior Scientist of Reproduction Biology, Kazak Institute of Sheep Breeding. Research on relationship between sheep productivity and new management regimes under privatisation in Kazakstan; 8 publications, including some in English.

E.J. Milner-Gulland Renewable Resources Assessment Group, T.H. Huxley School of Environment, Earth Sciences and Engineering, Imperial College London.

Sarah Robinson Research Assistant, University of Warwick/EU project on Land Degradation and Agricultural Change in Kazakstan. Research from 1997–8 on biological and socio-economic aspects of the privatisation of state livestock farms.

Ogultach Soyunova Director, Department of Privatisation in Agriculture in Ministry of Agriculture, Turkmenistan. Research on changing economic investments and returns from livestock farming. Research specialisation: changing household economic structure of pastoralists in the reform period; 10 publications, including some in English.

Hélène Vidon Centre de Coopération Internationale en Recherche Agronomique pour le Développement, Montpellier.

Iain A. Wright Manager of Ruminant Resource Use Programme, Macaulay Land Use Research Institute, Aberdeen. Ruminant nutritionist with a research background in optimising feed balances between sown and natural resources to meet both production and conservation objectives, and evaluating policy impacts on livestock systems.

PREFACE

This book documents some of the impacts of transition from state to private management of livestock and pasture land in two Central Asian countries – Kazakstan and Turkmenistan. The findings are based on collaborative research between a group of natural and social scientists over a two-year period from 1997 to 1999. The authors are British, Turkmen, Kazak, American and French, representing a range of disciplines including social anthropologists, animal scientists, pasture agronomists, a veterinarian, climate ecologist and an economist. The research was funded by the Department for International Development[1] of the British government, with the exception of research carried out by Sarah Robinson and E.J. Milner-Gulland (Chapter 7) which was funded by the INTAS programme under the European Union.

1 The United Kingdom Department for International Development (DFID) supports policies, programmes and projects to promote international development. DFID provided funds for this study as part of that objective, but the views and opinions expressed are those of the authors alone.

ACKNOWLEDGEMENTS

Gratitude is due to the many individuals who contributed to the research. First among these must be our interpreters, who painstakingly and skilfully conveyed meanings between researchers who did not speak each other's languages, and between researchers and the citizens of the two countries. We thank: Jabbar Abdul, Timur Ahmedjanov, Abilai Karibaev, Nurlan Malmakov, Kamshat Moldebekova, Ata Murat, Yerjan Nurhasanov, Roza Nurjanova, Timothy Okraev, Sayat Temirbekov and Gennady Ukhanov.

Encouragement and assistance was received from Shirin Akiner, Kasim Asanov, Agajan Babaev, Roger Blench, Allen Brown, Angela Croucher, Mima Garland, Gustave Gintzburger, Stuart Horsman, Cara Kerven, Emilio Laca, Simon Maxwell, John Milne, Mark Perkins, Chris Pescud, Jicksinbai Sisatov, Vladimir Yurchenko and Euan Thomson.

We appreciate the assistance provided by the British Embassies in both Kazakstan and Turkmenistan as well as support from the Institute of Deserts, Flora and Fauna and the Institute of Animal Husbandry (Turkmenistan) and the Institute of Pasture and Fodder (Kazakstan).

Lastly, we express our sincere appreciation to the many shepherds, officials, scientists and business people who willingly answered our questions and provided us with the hospitality that makes working in Central Asia always a pleasure.

NOTE ON CURRENCY EXCHANGE RATES

79 Kazak tenge were equivalent to one US dollar in 1998. This rose to around 140 tenge to the dollar in 1999.

5,000 Turkmen manat were equivalent to one US dollar in 1998. This rose to around 19,000 manat to the dollar in 1999, and 21,000 in 2000.

1

'WE HAVE SEEN TWO WORLDS'[1]

Impacts of privatisation on people, land and livestock

Carol Kerven

> *The steppe is wide as a sea*
> *It's decorated by flowers*
> *It lays beneath people's feet;*
> *So, be fertile, be people's joy*
> *We call you our Mother*
> S. Seifullin, Kazak poet
> (Translated from Russian by Yerjan Noorhasenov
> (in Asanov *et al.* 1992b))

Past and present

The lives of Central Asian pastoralists were engulfed by two different worlds in the twentieth century. First they were drawn into large collective farms, in which family animals became state property, decisions on animal husbandry were made by officials and technocrats, and hard work was rewarded not by seeing one's flocks grow to be passed onto one's children, but with medals and citations. After a period of great hardship, life in the collectives became secure. A decent house, income and health care was provided. Shepherds' children could be educated in the new village schools and go on to become university graduates. The threat of livestock being decimated by climatic disasters was overcome.

But then came another world, in which the previous order was turned upside down. An older shepherd reflects that:

> In Soviet times everything in our lives was settled. Now, who can help us? One will not help another if one does not help himself. If help does not come from foreign countries, it will not come from our government. After the break-up of the Soviet Union, no one evaluated how precious

---

1 Livestock worker on a former state collective farm in the desert, Almaty *oblast*, Kazakstan.

1

our livestock are. No one had pity about the animals. Now they have started to think about it but it's too late to increase them.

Nomadic pastoralism has an ancient heritage in the vast natural pastures of the Asian steppes. Even a century ago, 'There were yurts from central Anatolia to eastern Mongolia – a quarter of the way around the world, in a band 1000 miles from north to south' (Ryder 1983: 265).

The yurts, symbols of a nomadic way of life, still exist in Central Asia but the way of life they represent is fast unravelling. Major upheavals have profoundly altered the pastoral systems of Central Asia this century. After incorporation into the Soviet Union, pastoralists were collectivised into state farms, building an industrialised nomadism that was highly productive. Despite many radical changes, Soviet administration preserved some important elements of the nomadic system.

Independence from the Soviet Union brought new policies that dismantled or reorganised the state farms, withdrawing most state support, and decontrolled markets. This has been termed 'the second Revolution'. As pastoralists now confront a market system only the fittest may survive. Many have lost their livelihoods and more are at risk of doing so. Yet in the neighbouring ex-Russian ruled Republic of Mongolia, pastoralists are prospering by some accounts (Fernadez-Gimenez 1999; Sneath 2000). The same applies in the rangelands of western China (Benson and Svanberg 1998; Hamann 1999). Why has reform in the former Soviet Republics of Central Asia not benefited the remote dwellers of the deserts and steppes? This book attempts to provide some answers to this question.

Privatisation policies have proceeded differently in each country. Policy changes in the livestock sector of Turkmenistan have been neither as profound nor as sudden as occurred in Kazakstan, and the results so far are a greater degree of stability in the livelihoods of Turkmen pastoralists. The Turkmen government is following a 'step-by-step' approach to restructuring the economy, including the livestock sector. While frustrating to Western donors and investors, on the ground this approach seems to hold certain advantages for livestock producers, who are still somewhat protected from negative market forces while having an opportunity to accumulate private livestock leased from the state. Unfortunately, the very success of this approach is undermining state resources, and some reversals of policy are taking place.

In contrast, pastoralists in Kazakstan have been subjected to radical changes in their production system over a very short period of time. Reformulating the entire institutional, property rights and economic context of production has created opportunities for a few individuals to prosper, but has left many vulnerable to the cupidity of newly-empowered officials and entrepreneurs, with little economic support or protection from the state.

In both countries, the reorganisation of former state collective farms into a variety of new institutional forms, together with the introduction of open markets, has altered the goals and methods of livestock production. As a result, economic returns, the systems of natural resource management, and ultimately the welfare and income of producers are all affected.

The pastoralists of Central Asia are a human resource for the future development of the natural pastures and livestock industry. They possess knowledge and skills of irreplaceable value for managing livestock in a variable and challenging environment. They can make use of a great landmass that would otherwise remain unproductive. They can supply high-quality livestock commodities much in demand domestically and with a comparative advantage internationally. Under the right conditions, they can serve as guardians of the biodiversity in the land. Without these options, the people formerly dependent on livestock must leave the land to join the unemployed and unskilled in the cities.

The ecological foundations

The impacts of policy and economic shifts have been modulated by the ecologies of the rangelands in each country. There are substantial climatic differences between the northern and southern pasture regions of Central Asia which result in dissimilar natural limitations for livestock production in Turkmenistan and Kazakstan (Babaev and Orlovsky 1985; Khazanov 1984).

The northern desert and semi-steppe zones cover the middle part of Kazakstan in which this study was carried out. These zones experience a sharp continental climate with extremely cold and snowy winters during which plant production ceases for several months. To find grazing for their animals, pastoralists in the past had to move sometimes long distances before winter either to warmer southern desert zones or sheltered areas. Precipitation occurs mainly in two peaks during spring and autumn, with a gradient from 200–400 mm per year generally decreasing from north to south.

The southern desert zone, which includes most of Turkmenistan, has milder winters with moderate frosts and infrequent snow. The growth of vegetation is only interrupted for a maximum of ten days over winter and animals can thus graze most of the year. Rainfall is very sparse, at around 100 mm/year mostly falling in winter. Summers are extremely hot, with temperatures up to 50°C causing high rates of evaporation.

For pastoralism, the difference between the northern and southern regions is that the factors of cold and snow mainly shape seasonal livestock movements in central Kazakstan, while seasonal livestock movements in Turkmenistan are determined principally by availability of water.

Soviet planners addressed these limitations in each case through the application of science and engineering. Water was provided to the desert, and fodder to the animals over winter. As the post-Soviet states retreat from maintaining services, these environmental barriers to production have re-emerged.

3

Kazakstan: sudden dissolution of state farms

Historically, Kazakstan supported a nomadic pastoral production system. Long distance migrations, a response to strong seasonal variations in climate and range production, were a critical part of the ecological adaptation of Kazak pastoralists (Chapters 3 and 7). Under the Soviet system imposed in the 1930s, nomadism was curtailed. People and livestock were forcibly sedentarised into a collective structure of state-controlled farms (Zveriakov 1932). While some limited seasonal movement continued, the emphasis was on intensifying livestock production. Intensification required state investment in local infrastructure, both for livestock and people, through provision of fodder, water, transportation and infrastructure. Between the 1950s and 1980s, this was accomplished through increasing the livestock fodder base, reversing the traditional pattern of extensive utilisation of natural pastures, as documented in Chapter 3. The result was reduction of mobility and long-range movements, and greater grazing pressure on local pastures. This strategy was implemented throughout the rangelands of the former USSR, not just in Kazakstan, and it was maintained by state subsidies on fodder, livestock transportation, and village-centered facilities for pastoralists.

Independence for Kazakstan was followed by a massive restructuring of the economy, away from central planning and toward an open market system. Shortly thereafter, the Kazakstan livestock sector collapsed, ostensibly as a result of economic restructuring and privatisation. Evidence in Chapters 3 and 5 suggests however, that the range livestock system may have already been in a precarious state prior to independence. Sixty years of Soviet management had apparently taken its toll on Kazakstan's rangelands (Babaev and Kharin 1992; Gilmanov 1995), while also reorganising pastoralist livelihoods in such a way that people became completely dependent upon state inputs of livestock feed, fuel, food and other commodities, rather than on local resources.

The decollectivisation of state farms and a move toward partial privatisation of land and livestock holdings began in 1994. The sequence of events and their immediate effects is given in Chapter 5. At the same time as the assets of farms were privatised, the state withdrew from both the provision of subsidised inputs and the buying of produce. This sudden cessation of state support had an enormous impact on the livestock sector. The newly-formed cooperatives inherited debts and could not afford to buy inputs (seeds, fertiliser, feed, veterinary supplies, fodder, fuel, spare parts) while marketing channels were not in place to substitute for the former state-controlled system. The cooperatives could not pay their members or creditors and started to do so by the distribution of goods, especially livestock and farm machinery. This further reduced the assets of the cooperatives, to the extent that most are no longer viable. Rural areas reverted to a barter economy.

The national livestock population fell precipitously following the initiation of privatisation in 1993–4. This rapid decline continued through 1996 with further losses and then slowed somewhat through 1997 and 1998. Overall, the figures charted in Chapter 5 indicate a total decline of 70 per cent of the national small stock population (from roughly 33 million in 1993 to only about 10 million in 1998). A variety of factors, including the loss of markets, retraction of fodder subsidies and the move from collectivisation toward privatisation contributed to this rapid and massive destocking. But it is likely that unsustainable land use practices and a significant shift in climate patterns in the 1990s also contributed to the demise of the livestock system.

Chapter 4 shows that rangeland degradation and increasing climatic variability set the stage for the collapse of the livestock sector. The economic transition then suspended the flows of fodder and other inputs, leaving the pastoralists in deep economic difficulties, and with a seriously depleted environmental resource base. Just how depleted is demonstrated in Chapter 4. Particular pasture zones were subject to severe decline in condition or major degradation over several decades. These zones are now showing strong signs of recovery following the release of grazing pressure due to destocking. A similar pattern is portrayed in Chapter 3.

The financial crisis and destocking occasioned by privatisation have made seasonal livestock migration either impossible or unnecessary for reasons discussed in Chapters 5 and 7. The number of animals owned is frequently too low to justify the cost of moving, transport is lacking and roads to distant mountain pastures are no longer maintained. Families can no longer rely on the services formerly provided by the state farms to mobile herders, and now do not wish to move away from village service centres. Stock theft and lack of sufficient labour are other factors discouraging migration. Moreover, the extent of destocking is such that more accessible pastures are often underused so there is no need to move elsewhere.

In the past the livestock sector played a major role in the economic activity of the country. Kazakstan produced almost a quarter of the Soviet Union's lamb and one-fifth of its wool (McCauley 1994). In the Soviet era trade with former members of the USSR accounted for nearly 90 per cent of livestock exports from Kazakstan prior to 1993 (Kerven *et al.* 1996). Many of the sheep production systems were geared towards wool production to satisfy the huge demand for wool in the rest of the Soviet Union (Chapter 6). The break-up of the Soviet Union effectively cut off this market for wool. Alternative markets were not immediately available, as world prices for wool slumped in the early 1990s. The loss of these markets helped to undermine the economic viability of the Kazakstan livestock sector.

Livestock productivity has been compromised due to the loss of state inputs. Many newly privatised pastoralists cannot afford to provide adequate nutrition for their few animals, either with supplementary feed or by moving seasonally to more nutritious pastures. Under climatic conditions of long, cold winters, supplying animal feed over winter has again become acutely problematic, as noted in Chapter 6. The result is that some animals have a lowered output, which further reduces the economic viability of small family livestock enterprises. However, new private farmers who have greater resource endowments are able to overcome this constraint, through more successful engagement with markets that have recently developed (Chapter 8).

The speed of change and degree of dislocation has left many losers but some winners. Chapters 7 and 8 give some indication of growing socio-economic differentiation among pastoralists, partially in response to the market but also as a direct effect of decollectivisation. After an initial period of shock, by the late 1990s marketing systems had spontaneously developed to meet urban demand for meat, in spite of neglect by the government. A small proportion of newly privatised livestock owners and an emerging group of traders have responded to commercial incentives. Marketing is now firmly in the hands of private entrepreneurs and the state no longer has a role. But the majority of pastoralists are unable to benefit from the market, as their private flocks are too small or they live at too great a distance from urban centres. For such people, animals can provide an important source of subsistence but with a low and insecure cash income.

Several themes recur from the evidence on Kazakstan presented in this book. The seasonally and spatially variable pastures provide a valuable natural resource base for the livestock industry and the livelihoods of people. In the latter part of the Soviet era this resource was pushed beyond its limits to meet unsustainable production targets. Some parts of the rangelands have the capacity to recover and are now showing signs of regeneration following a huge decline in stocking pressure. However, an extreme climate will inevitably necessitate some dependence on supplementary winter feed. This has become the principal impediment for the majority of pastoralists. Under the new market economy, only a few livestock-owners are able to provide adequate animal nutrition and market their animals successfully. Through reinvestment, these people are accumulating livestock and profiting from the pastures, at present.

There are several risks for the future. Larger-scale livestock-owners may squeeze out poorer pastoralists by claiming the best natural resources. Pasture land around settlements is now being overused while distant pastures are barely grazed. This could cause further degradation around key points. Whether the new livestock enterprises resume the seasonal movement of their animals in future depends on national land tenure policies as well as on market signals. It will be important to allow relatively open access to seasonal pastures that would give more pastoralists the option of migrating again. Pastoralists are likely to resume moving their animals by season when costs of moving are outweighed by benefits from a revitalised livestock economy.

A matrix of conditions will be necessary to ensure the future of pastoralism in Kazakstan. Land policies must allow livestock movement in order to preserve and make the most efficient use of the natural resources. Some rehabilitation of a winter fodder base will be necessary, requiring the deployment of national scientific expertise as well as updated technical methods suitable for a market economy. Lastly, the revival of a livestock industry that can absorb more people will depend upon the expansion of remunerative markets, both domestically and internationally. The state can play a more positive and enabling role in this effort, but ultimately it will be the commercial sector that takes the lead.

Turkmenistan: the gradualist approach

The Turkmen government has explicitly adopted a gradualist approach to privatisation. Whilst policy decisions have produced profound changes in the rangeland production systems, the process has proceeded in a series of steps, with consequently fewer sharp shocks compared to the rapid pace adopted in Kazakstan. In particular, the huge destocking experienced by Kazakstan has so far been avoided. Moreover, the tighter controls exerted on state farm directors by the state is in marked contrast to the power accumulated by ex-state farm directors in Kazakstan as a result of privatisation.

Livestock management has largely been transferred from the state to leasing associations and private individuals. This process and its impacts are described in Chapter 9. The associations are still administered by the previous management. Animals are leased to shepherds who are responsible for their management. The annual produce of the herd is divided between the leaseholder and the farmers' association using contracts that vary by duration and location. Pasture land and state farm infrastructures have not yet been leased to individuals, another difference with Kazakstan.

Leasing provides opportunities for formerly-employed shepherds to build up private livestock holdings, described in Chapter 12. A transfer of livestock assets from state to private hands is thus taking place. However, associations are having difficulty covering their overhead costs (mainly salaries for management) from fewer numbers of animals. Farmers' associations were initially responsible for providing their shepherds with inputs and services at cost price, but increasing financial pressure is reducing their ability to assist members. Most new associations have huge deficits as they have inherited dysfunctional administrative structures.

Association members and private shepherds alike are having to rely on their own resources. Economic realities are accelerating the pace of the official 'step by step' approach to privatization, as withdrawal of government support pushes people to find private solutions quickly. More capable and entrepreneurial shepherds are adopting strategies to cut costs and increase returns, such as limiting fodder costs or reducing long-distance seasonal movements (Chapter 11). Shepherds are now obtaining inputs privately and have almost complete control in the way they manage their animals and use their environment. For some pastoralists, the leasing system has been successful, while others are finding that leasing is not economic.

The winter climate of Turkmenistan is less severe than that of Kazakstan. This leads to an important difference for livestock management. Sheep and camels in the Kara Kum desert can graze on pastures year-round, only needing supplementary feed in rare severe weather or for weak animals. Privately managed livestock receive better feeding than association livestock. Purchase of fodder is the main expense on livestock among owners of private animals (Chapter 12). Owners of small flocks spend more on purchasing fodder than do owners of large flocks who use more productive remoter pastures (Chapters 11 and 12). But removal of state support increases the vulnerability of pastoralists to climatic shocks; by affecting pasture productivity such events could cause large-scale loss of animals in the future.

The new associations cannot afford to repair the waterpoints that were formerly maintained by the state. Access to water, the cost of obtaining water and its quality are becoming the main limiting factors for livestock production. The implications of new patterns of water use are assessed in Chapter 10. Water quality in much of the desert is poor, with high salt concentrations. Patterns of movement hinge around water. Reduced state support results in seasonal restrictions on good water supply that leads to under-use of some distant pastures and overuse of pastures around the better water points. This has caused and will continue to cause long-term negative changes in the vegetation around some water points.

There have emerged three interest groups with different rights to land, water resources and livestock:

• Private shepherds with their own animals and formal access to village wells only.
• Hired private shepherds who have undertaken to look after other peoples' animals at a monthly rate. They have the same rights over water and land as private shepherds.
• Farm association members, who are able to use their association's pastures, wells and other water resources for leased animals as well as for their own private animals.

While each group has different formal rights, in reality resource use is more fluid. In another contrast to Kazakstan, there has been much less informal privatisation of landed property resources and key resources can still be shared communally (Chapter 9).

Seasonal livestock movement is reduced, for similar reasons as found in Kazakstan. In the Soviet period, many of the state-owned sheep were moved to northern pastures in the winter. As the new associations withdraw support, fewer pastoralists are undertaking the winter migration as the costs incurred make it unattractive. Instead they rely on village wells and surrounding pastures for longer periods, increasing the grazing pressure on these areas while northern pastures may be underused (Chapters 9 and 10).

Loss of state support and employment is leading to growing economic differentiation between pastoralists, shown in Chapters 12 and 13. Due to the extremely arid environment, desert-dwellers are unable to diversify into crop farming if dependence on animals becomes less possible. Poorer families have moved away to farm in the irrigated zone, and livestock in the desert are being concentrated in the hands of the remaining better-off pastoralists or managed by hired shepherds.

Market privatisation has been initiated but the state remains involved in marketing, which is another difference with Kazakstan. As the proportion of privately owned animals has been increasing while that of state animals declines, the main flow of animals and products to the market is now through non-state channels. In Chapter 13, the parallel co-existence of state and private marketing channels is discussed.

Animal sales by the farm association administration are not on a financially sound basis. The new leasing system means that many pastoralists no longer receive a salary from the state and animal sales now provide the major source of cash income, mainly required to buy food. Traders have started to buy and sell livestock since independence, while some state livestock processing plants have been semi-privatised under new leasing arrangements. Private capital is also being invested in wool and pelt marketing.

The instability of Turkmenistan's economy continues to rock the new organisational structures constructed on the ruins of the former state farms. Fluctuating prices and high costs of production make it extremely difficult for the associations and pastoralists to align their management strategies with the new demands of a market-based system. In 1998 they experienced the shock waves of Russia's economic crises which halved the value of the national currency, the Manat, in three months.

The state has maintained a controlling role in monitoring agricultural production; targets must be met and plans fulfilled. The government has a policy of food security to be achieved mainly by increasing wheat production. This has happened at the expense of feed crops for animal production, as irrigated land is limited (Chapter 11). There is also a state plan to increase the national sheep population each year.

An unpredictable economy, coupled with government demands to achieve certain production targets, restricts the ability of associations to become successful entities. Institutional structures are characterised by Soviet-style hierarchies and a rapid turnover of personnel in positions of responsibility. This encourages distortions and is hindering adaptation to the market. So far reforms have focused on re-allocation of livestock and their returns, rather than addressing the structure of the institutions themselves.

The immediate withdrawal of the state from the production side of the livestock sector would create further problems, as has been evident in Kazakstan. The minimal amount of state support still given to associations at least allows these to gradually hand over animals to pastoralists through the leasing system. Pastoralists can then experiment for several years to find the most cost-effective methods of private livestock management. If the associations were to be suddenly dismantled and their assets distributed, the many shepherds left with small flocks could not make a viable livelihood given present market prices, especially for wool. The result – looking at what has happened in Kazakstan – would be a sharp decline in animal numbers and a concentration of remaining animals into fewer hands. Unemployment and meat prices would rise, as fewer people would be able to raise fewer animals.

2

AGRARIAN REFORM AND PRIVATISATION IN THE WIDER ASIAN REGION

Comparison with Central Asia

Carol Kerven

Introduction

Reforms over the past decade in Central Asia have generally had negative impacts on the systems of extensive pastoral livestock husbandry which are the subject of this book. To more fully understand why this is so, it is useful to compare the situation in Kazakstan and Turkmenistan with other ex-Soviet Central Asian states. Expanding the comparison wider, it is found that the livestock sector has fared better in other Asian states which are transforming from command to market economies but have undergone rather different reform processes in the past decade.

We begin by considering the general trends of agrarian reform in Central Asian countries, considering how the chronological sequence of key reforms and the speed with which they have been implemented has affected the pastoral livestock sector in each case. The privatisation of livestock and rangelands is the central feature of the reform process in this sector, moving from a socialist ideology of state property to that of individual ownership or control. The analysis then shifts to the wider context within which reform agendas have been undertaken in Central Asia, with a general assessment of the impacts of reform. Lastly, comparisons are drawn with recent agrarian reform in southeast Asia, China and Mongolia.

Pace and sequencing of reform

Agricultural reform in Central Asia following the break-up of the Soviet Union has proceeded differently in each country. Reforms have also been implemented at different speeds, with Kyrgyzstan beginning the process earliest, followed by Kazakstan, both adopting simultaneous reform packages on the advice of international financial institutions. Uzbekistan and Turkmenistan are following a

10

sequential approach to introducing reforms, while reforms were delayed in Tajikistan due to civil war (Pomfret 1998; 1999).

Given that each newly-independent Central Asian state has pursued a different timetable for agrarian reform, inter-country comparisons at any one point in time may be misleading. By the mid 1990s, five years into the transition period, Pomfret noted that output in Central Asia had followed a U-shaped curve; slower reformers such as Uzbekistan and Turkmenistan were still on the downward part of the U while the early fast reformers were beginning to climb up on the far side (Pomfret 1998). By the end of the decade, at least for agricultural output, this picture was more complicated.

A further difficulty in assessing agrarian reform in Central Asia is that the process and impacts are not well-documented, particularly for Turkmenistan and Tajikistan. In general, quantitative measures of economic growth or decline in Central Asia are subject to some scepticism, based as they usually are on official statistics. The collapse of the old economic order has engendered a widespread increase of informal economic activities. Extended families, struggling to make a living, exchange goods and services outside of formal channels and do not save cash in the unreliable banking systems. Large organisations sometimes still conduct their transactions by barter, at unrecorded rates of exchange. The flourishing cross-border trade is not always accurately documented. Everyone tries to avoid paying tax on production, income and exports. The extent of the informal economy is such that agricultural output and trade is very likely to be underestimated in official figures (IMF 1999).

Kazakstan

One of the earliest reformers in Central Asia, the new Government of Kazakstan attempted to quickly convert to a market economy with considerable encouragement from international organisations (mainly through the Bretton Woods group). These organisations offered financial loans and grants conditional on adopting the package of structural and market reforms which had been implemented in Latin America (Kalyuzhnova 1998; Pomfret 1998). The legal framework was enacted in a relatively short time, from 1992 to 1996, covering land distribution, state farm privatisation, removal of state prices, cancelling of state purchase orders, and liberalisation of agricultural trade. These changes were implemented during the difficult early period of disengagement from the Soviet Union economy, which entailed currency revaluations, loss of subventions from the centre, and loss of the captive markets for agricultural products within the Union (Spoor 1996).

Several problems for livestock production ensued from the unfortunate conjunction of simultaneous legal changes and economic crisis. State financial support was removed, leaving producers without access to agricultural inputs. State livestock farms in some regions of Kazakstan had been heavily dependent on subsidies and inputs imported from other Soviet countries (World Bank 1993).

Since the removal of state support coincided with a severe currency crisis and hyperinflation, the new agricultural entities were obliged to barter their most exchangeable assets, which were principally smallstock, at unfavourable terms of trade for fuel and basic supplies. Second, laws created by the top level of government have been variously interpreted or subverted at the regional and local administrative levels (Spoor 1996; Ye 1999).

This has led to the maldistribution of state farm assets, including land, in many instances (ADB 1996a). Third, state distribution channels were abolished before commercial markets had evolved, at the same time as the former Russian outlets for livestock products closed down. While the price of meat had dropped due to oversupply in the first period of economic crisis, rising urban incomes in the second half of the decade led to an increase in meat prices. But loss of the former Soviet market has continued to depress the price of the other livestock products of wool and pelts. Domestic processing plants closed down due to lack of government finance and large debts, while international demand for wool was weak throughout the 1990s.

The impacts of swift and deep legal reform in Kazakstan caused economic and social disruptions in each sector, including the extensive livestock sector. In the first few years after independence grain farming likewise experienced a severe contraction of output to one-third of the previous level, though this trend seems to be improving (see Figure 2.1). Overall, agriculture suffered from the general effect that: 'On the one hand the new economic mechanisms were not constructed, whilst on the other hand the centrally planned rules were instantly abolished' (Kalyuzhnova 1998: 155).

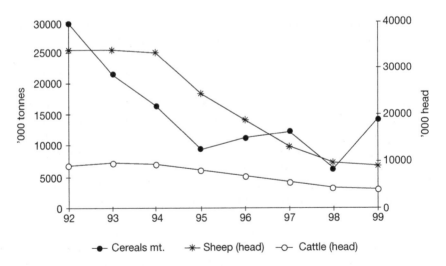

Figure 2.1 Kazakstan: cereal and livestock production, 1992–9.

Source: FAO 2000.

Kyrgyzstan

The agrarian reform programme has been the most vigorous and far-reaching of all the central Asian states, with continual encouragement and inputs from the major international financial bodies (IMF 1999; Röhm 1995). The livestock sector had been relatively more important to the national economy of Kyrgyzstan than in any other Soviet Central Asian state, contributing one third of gross domestic product and over 40 per cent of agricultural output (World Bank 1993). Due to intensification programmes, a high livestock population was dependent on mainly imported feed over the winter period (van Veen 1995). The livestock sector was particularly hard hit by the restructuring of the economy combined with the loss of Soviet subsidies and markets. This caused a near-collapse, as occurred in Kazakstan for similar reasons (see Figure 2.2). Between 1990 and 1996, nearly two-thirds of the national sheep population and one-third of the cattle had disappeared. The rate of loss then stabilised. The livestock sector remains depressed while food crop output has improved markedly, as smallholders try to meet subsistence needs and urban demand, as shown in Figure 2.2 (ULG 1997; World Bank 1997).

Privatisation of state property, including farm land and livestock, was an early goal largely achieved in the case of livestock by the middle of the 1990s, although with great losses of animals involved and ensuing rural poverty (Howell 1995). Pastureland has not been privatised, but is controlled by local authorities and used on a leasehold fee-paying basis (Bloch and Rasmussen 1997). Donor-funded programmes targeted at the livestock sector in the latter part of the decade have yet to have an impact on livestock numbers.

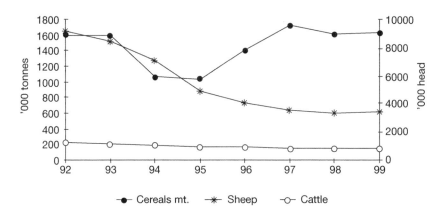

Figure 2.2 Kyrgyzstan: cereal and livestock production, 1992–9.

Source: FAO 2000.

Turkmenistan

During the first half of the independence decade, reform in Turkmenistan and thus in the livestock sector was deliberately limited in order to introduce phased and regulated market conditions (Khalova and Orazov 1999). Agrarian reforms began in 1995 by a process of reorganising state farms into peasant associations. This has involved the gradual reduction of state financial and technical support to farms, the transfer of livestock from state to private ownership through leasing, creation of small irrigated crop holdings with individual title, and partial liberalisation of marketing and price controls for agricultural commodities (Ataev 1999). A prominent positive result of this slower pace of reform has been avoidance of a drop in livestock numbers, compared to Kazakstan and Krygyzstan (see Figure 2.3). Livestock numbers have remained relatively stable throughout the 1990s, though shifting from the state to more than half now in private ownership. Although official figures are not necessarily reliable, these patterns are confirmed by knowledgeable officials.

There have been major negative effects of government disinvestment in the livestock sector; reduction in the quality of wool and pelts, non-maintenance of key facilities such as water points, and a shortage of fodder (documented in Chapters 9, 11, 12 and 13). Political emphasis on wheat production for self-sufficiency and cotton for export revenues has meant that, after a steep drop in the mid 1990s, output has since risen, with wheat overtaking cotton (Figure 2.3). However, some irrigated land previously assigned to fodder crops has been converted to wheat production, raising the price of purchasing fodder for pastoralists.

One of the now-recognised ingredients for successful transition to market economies has also been lacking: the promotion of new enterprises and open trade policies (Pomfret 1999). State intervention in pricing mechanisms has also reduced

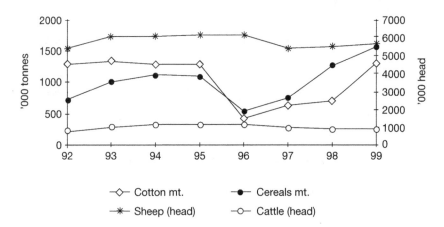

Figure 2.3 Turkmenistan: cotton, cereal and livestock production, 1992–9.
Source: FAO 2000.

real incomes for producers and incentives for exporting livestock products (Pastor and van Rooden 2000).

Uzbekistan

Agrarian reform in Uzbekistan has had some similar goals and methods to that in Turkmenistan, with disappointing outcomes in terms of total agricultural growth (Pomfret 2000a). This has been mainly due to powerful sectoral interests supported by a government that relies heavily on agricultural revenue from cotton. Since 1992, agricultural output has declined only slightly in some sub-sectors, much less than in the other Central Asian countries (see Figure 2.4). Output increased for cereals. Decollectivisation has not been total, as old state farms remain in a new guise, and responsive to upper levels of government control and demands for cotton output. Livestock output remained more or less steady during 1991–7, but has declined in more recent years though this is not reflected in official statistics (Zanca 1999). Reasons for this decline are undoubtedly related to the reduction in state support to agricultural enterprises. Newly privatised livestock farmers are relatively independent from state control, but rangeland has not been privatised. Four-fifths of livestock were privately owned by 1998 and private holdings supplied over two-thirds of animal products to the market. Successful pressure by government to increase wheat output has pushed out fodder production, reducing livestock productivity.

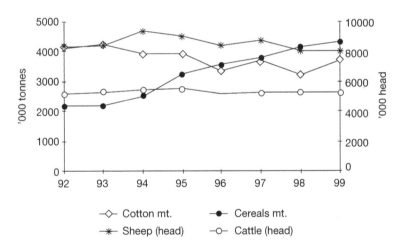

Figure 2.4 Uzbekistan: cotton, cereal and livestock production, 1992–9.

Source: FAO 2000.

Context and consequences of agrarian reform in Central Asia

Agrarian reform in Central Asia was undertaken within wider agendas of national economic programmes and international pressure for structural readjustment. After ten years, it is possible to assess some of the consequences on agricultural output of these differing agendas.

International influence and development assistance

Following independence from the Soviet Union, multilateral lending organisations such as the International Monetary Fund (IMF), Asian Development Bank (ADB) and the World Bank offered the new republics loans conditional on initial speedy implementation of structural reform measures. These measures included and also affected the agricultural sector. As already noted, Kazakstan and Kyrgyzstan more readily accepted rapid reform, and have received much higher levels of credit than Turkmenistan and Uzbekistan from the multilateral organisations (Table 2.1 and Figure 2.5). However, questions are now raised as to whether applying the prescribed pillars of transition, comprising privatisation, market liberalisation and state withdrawal, has necessarily led to economic growth (Spoor 1997a).

Differences in the pattern of foreign financial involvement in each Central Asian republic also reflect their inherent economic potential. Private lending to Kazakstan, Turkmenistan and Uzbekistan has clearly been bolstered by those countries' capacity to generate future foreign earnings from oil, gas, minerals and cash crops. Commercial confidence in Kazakstan, expressed as direct foreign investment, is particularly noteworthy. Kyrgyzstan, with few economically-valuable resources, has relied heavily on the multilateral development agencies for credit. The present value of Kyrgyzstan's debt is nearly half the country's gross domestic product.

Impacts of reform programmes on agricultural output

Agricultural performance in the first ten years of transition does not appear to be linked to early adoption of the prescribed structural reform packages and associated loans. In Kazakstan, agriculture's contribution to the economy has shrunk in relative and absolute terms since 1990 (Figures 2.1 and 2.5). Private credit and investment has largely shunned the agricultural sector, concentrating instead on services and energy. Kyrgyzstan, on the contrary, has been forced to increase food output with the loss of Soviet subventions and lack of foreign exchange to import food. Grain output has risen although it is not clear whether development credits have been responsible (Figures 2.2 and 2.5). In neither Kazakstan and Kyrgyzstan has the livestock sector bounced back from the early shocks of reform.

Agricultural performance has been, in fact, better in the two countries which have not undertaken fundamental structural reform on the IMF/World Bank

Table 2.1 Central Asian selected economic indicators 1998

Country	Present value debt/GDP (%)	Per capita debt present value (US dollars)	Debt to multilateral financial organisations (%)	GNP per capita (US dollars)	GDP average annual growth	Agric. annual average growth	Agric. % of GDP 1990	Agric. % of GDP 1998	Direct foreign investment per capita (US dollars)
Kazakhstan	24.5	364	42	1,310	−2.5	−18.9	27	9	73
Kyrgyzstan	48.6	333	73	350	1.8	6.0	34	46	23
Turkmenistan	90.2	427	0.05	n/a	5	n/a	32	25	26
Uzbekistan	14.8	138	14	870	3.4	3.0	33	31	8

Source: World Bank 2000.

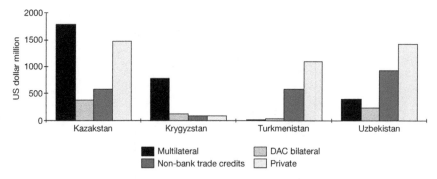

Figure 2.5 Accumulated international credit to four Central Asian republics by 1999.

Source: Joint BIS-IMF-OECD World Bank 2000.

Notes

1   Multilateral claims include the Asian Development Bank (ADB), IDA credits from the World Bank, the International Monetary Fund (IMF) and the International Bank for Reconstruction and Development (IBRD).
2   DAC (Development Assistance Committee) bilateral loans from OECD countries.
3   Non-bank trade credits are official guaranteed export credits from OECD countries.
4   Private loans from banks in eighteen major industrialised countries and offshore centres.

model. The government of Turkmenistan, having received very little credit from the multilateral organisations, began in the mid 1990s to finance grain and cotton production through private foreign bank credit. This has resulted in rapid increase of these crops (Figures 2.3 and 2.5). However, Turkmenistan now has a high level of indebtedness partly due to large-scale prestige urban building projects. Uzbekistan has had the most stable agricultural performance and least reliance on foreign credit, whether private or multilateral. Agricultural output has been sustained by a relatively healthy economy, as Uzbekistan has been able to continue exporting cotton and gold for hard currency throughout the transition period.

For the livestock sectors, those countries (Kazakstan and Kyrgyzstan) which enacted rapid reform had a liquidation of livestock assets. This left large numbers of former state farm employees with no means of production in environments where other means of livelihood were not feasible. Price decontrol and removal of state orders raised the costs of production while output prices were depressed, leading to a cost–price squeeze for producers from which many never recovered and simply went out of business. In the confusion caused by simultaneous reform laws, uneven application and misappropriation served the interests of well-placed official elites. One result is the reappearance of wealth differentiation in stock ownership, as before the collectivisation exercises of the 1930s.

Paradoxically, reforms aimed at land privatisation, which was viewed as a springboard for agrarian transformation by encouraging rural investment, has hardly affected the extensive livestock sector. Very little rangeland, in contrast to irrigated cropland, has been actually privatised (by lease or outright ownership) in those countries where this has been permitted. Several commentators note that the reformers' emphasis on legal privatisation of agricultural land as necessary for growth was misplaced. Security of tenure can be achieved without formal transfer of ownership, while effective growth required commercialisation of land through access to credit and foreign exchange, which has been difficult to obtain in most Central Asian rural economies (Griffin 1999; Ye 1999). Moreover, where privatisation of rangeland has taken place, it has been on a small scale by wealthier livestock owners, who have sought to control key resources within rangelands rather than vast hectarages of pasture. Privatisation of state rangeland infrastructure, such as wells, barns and agricultural machinery, rather than land itself, has allowed a minority of livestock owners to commercialise their operations.

Market liberalisation and privatisation

Market liberalisation has been a predominant goal of the reform lobby, a goal broadly achieved in the livestock sectors of Central Asia but mainly confined to domestic markets. Price decontrol of livestock products – meat, milk, wool and pelts – was an early part of the post-Soviet legislation. Producers and incipient traders responded fairly quickly to rising consumer demand for meat and milk, as the state ceased to supply these goods and also as private marketing of these products had not been totally eradicated in the Soviet era. However, market access has become problematic for producers in remote rangelands due to transport costs and for entrepreneurs due to growing corruption in market transactions. Consequently, unequal participation in the new commercial markets, or participation at unfavourable terms, has been one of the main causes of rising poverty among livestock-keepers.

In those countries where privatisation of state industries had been an early component of reform, domestic demand for non-edible livestock products largely terminated as processing facilities were not successfully privatised. A crucial part of economic restructuring to a market economy was deemed to be the transfer of state industrial property to private ownership. But many state factories in the livestock sector (as in other sectors) were debt-ridden, large-scale centralised operations which were not financially viable after prices for inputs (e.g. fuel) rose and production declined. This was the case in Kazakstan and Kyrgyzstan where, however, foreign and national investors are now beginning to rehabilitate these processing factories.

Where the state still maintains a controlling share of the manufacturing facilities, as in Turkmenistan, lack of investment in modern processing equipment has lowered output capacity, while profitable new markets are not actively sought. In all cases, though, development of export markets for livestock products has been hampered by a number of factors, including high customs tariffs, excessive bureaucracy, unrealistic official export prices, and inability to meet international standards. Trade liberalisation therefore has been less successful than market liberalisation for the Central Asian livestock sectors.

The rapid implementation of privatisation measures is now considered in some quarters to have been mistaken. The United Nations Development programme has concluded that policies on economic reforms should have been introduced in sequence rather than simultaneously, together with active state intervention to promote investment and ensure equitable access to livelihoods and resources (UNDP 1999). Transition to a market system was interpreted as minimising the role of the state, which caused immense economic hardship for more vulnerable social groups. Institutional reform has also been largely neglected in favour of market-driven policies characterised by laissez-faire approaches (Griffin 1999). A close study of one Central Asian economy that underwent rapid reform led to the conclusion that:

> Taken as a whole, the evidence . . . suggests that the attempt at shock therapy (in economic terms) by Kazakstan was a mistake, and that a more gradualist approach (both by Kazakstan and the other Central Asian nations) may well have provided stability and greater returns in the long term.
>
> (Kalyuzhnova 1998: 154)

Comparison of slow and fast agrarian reform impacts between Turkmenistan and Kazakstan provides some validation of this critique. For one particular vulnerable group – pastoralists living in remote areas – rapid reform in Kazakstan has brought great economic hardship and inequitable access to resources, whereas this has still been avoided under the slower pace of reform pursued by Turkmenistan. In both countries institutional reform of the livestock state farms has been part of the transition plan but has been insufficiently supported or controlled, especially in Kazakstan. Investment in the livestock sub-sector, by direct state action and by encouraging private entrepreneurs, has been neglected in both countries. Instead, investment has been confined to the energy, manufacturing and crop agricultural sectors.

The social impacts of reform

The initial enthusiasm among Western financial and donor organisations for rapid and radical agrarian reform in the former Soviet Union has by now been tempered. Some of the unexpected consequences of the recommended structural reform programmes include widening gaps in wealth distribution and growing rural poverty – the social costs of readjustment.

There appears to have been a greater increase of poverty in those Central Asian countries whose governments adopted a fast pace of reform:

> The limited evidence on income equality and poverty suggests that the more rapid transition in the Kyrgyz Republic and in Kazakhstan has been associated with widening inequality and increased poverty to

a greater extent than the more gradual transition in Uzbekistan and Turkmenistan.

(Pomfret 1999: 61)

As there is inadequate data to show which sectors of the population are most affected, it is not clear whether the rural population dependent on livestock have experienced disproportionate loss of incomes. A study in Kyrgyzstan found that in the first period of crisis, rural families with livestock bartered or slaughtered these for food (Howell 1995). This occurred also in Kazakstan (see Chapter 5). This 'depleting strategy' could only fend off absolute poverty in the short term. Once private flocks and herds had been eliminated, such families had lost their source of subsistence. They left the pastoral sector to take up the very few options for income-generating activities or became dependent on kin. However, the livestock-keeping families of Turkmenistan and Uzbekistan did not experience a severe economic crisis stemming from rapid reform, and consequently many have been able to maintain their livestock holdings.

Measurements of poverty in Central Asia are difficult due to an absence of data and differing criteria (Falkingham *et al.* 1997). It is widely accepted that personal income alone is an insufficient criterion of poverty. Other measures include access to health and education facilities, as well as the quality of these facilities and the effect on life expectancy and educational attainment. Employment opportunities and the rate of unemployment are also used as measures of poverty, as is food insecurity. One global standard is the Human Development Index developed by the UNDP (UNDP 1999). However, this is weighted towards non-income factors such as adult literacy, on which ex-Soviet countries score highly due to previous investment in education. Studies show that educated urban employees in Central Asia have sometimes suffered a worse decline of real income than less-educated rural residents, due to loss of employment with urban-based state organisations, higher costs of living and less ability to obtain food from home production (Falkingham 1999).

A decline of state expenditure on social services by the state is another indicator of growing latent poverty. The proportion of gross domestic production (GDP) spent on health, education and welfare (pensions, child allowances, etc.) has fallen in all Central Asian republics since 1991 (Mehrotra 1999). Here, however, it is difficult to find any relationship with the pace and extent of reform in each country. For example, by 1997, social security spending as a proportion of GDP had fallen most sharply in Kazakstan and Turkmenistan (fast and slow reformers), while education spending was at the same rate in both countries. Kyrgyzstan, though a fast reformer, seemed to be increasing social service expenditures after a slump in the early 1990s.

Comparisons with agrarian reform in southeast Asia and China

The new Central Asian republics experienced economic depression in all sectors, including agriculture, during the initial reform decade following independence. For early reformers, the slump in the livestock sector was particularly devastating. By contrast, in the state-managed economies of China and southeast Asia, earlier reforms started more than two decades ago were followed by a rapid and positive supply response in agriculture. This sector led the way in the initial stages of reform and remains a strong contributor to the economy.

These differences in the performance of the agricultural sectors attendant on reform may be due to different initial conditions between the economies and their respective agricultural structures (Green and Vokes 1997). The Soviet organisation of agriculture was founded on heavy capital investment (roads, machinery, rural electrification, irrigation), large-scale production units, and transfers from the industrial to agricultural sectors. Agricultural goods were intended to supply a protected, continental-scale market. Fuel and other input prices were subsidised at below-world prices, and farms were over-staffed. Agricultural production was vertically organised and dependent on links between processing and distribution chains dispersed over a wide geographical area. When these links were broken with the collapse of the Soviet Union, 'a complex system, created over decades of central planning, was bereft of input supplies, markets and technical support' leading to the reduction in agricultural output (Green and Vokes 1997: 265).

In the first period of post-independence reform in Central Asia, terms of trade for agricultural outputs declined sharply relative to fuel, machinery (spare parts) and other inputs such as animal feed. As the capital-intensive agricultural was inherently inefficient and dependent on subsidies, when capital equipment was no longer available and transfers were withdrawn, the structure of agricultural production was revealed to be unprofitable. Consequently incomes in agriculture fell in real terms and relative to other sectors of the economy, notably small-scale trade and urban services, which expanded rapidly and drew labour away from agriculture.

In China and socialist countries of southeast Asia, reforms released peasant farms to become more efficient in their use of labour and land, and to respond to prices as these were gradually decontrolled (Green and Vokes 1997). Farming was dominated by paddy rice, for which there was a both a domestic and international demand during the initial period of reform in China and southeast Asia. Rice cultivation was based on small family plots, and had never been fully collectivised or state-controlled. The scale of rice production used a type of technology that did not have any particular advantages of larger production unit size, and was not vertically and horizontally dependent on other sectors of the economy.

This difference in scale of production and the technology used was fundamentally different from that of Soviet agriculture. Southeast Asian and Chinese farmers could respond rapidly and flexibly to reforms in pricing and trade liberalisation by investing more labour to increase output. A net impact of macro reform was a highly positive supply response from the agricultural sector. In contrast, the large, capital-intensive state farms of Central Asia with their financial debt, welfare and employment obligations, and dependence on inputs could not adjust to new economic conditions, and output fell.

Pomfret (1997) and Griffin (1999) argue that the disparity in economic performance between China, southeast Asia and ex-Soviet Central Asia is due to their differing goals of reform. The new governments in the former Soviet countries set out to change from centrally planned command to market economies. Some have succeeded in creating market economies but with less success in economic growth, as important preconditions for growth – such as small- and medium-scale enterprises – have not been promoted. On the other hand, the governments of post-Mao China sought from 1978 to stimulate economic growth, beginning with the impoverished peasantry that made up four-fifths of the population. This goal has been achieved. The principal strategy pursued by the Chinese has been export growth and encouragement of investment, while state control over prices and markets has been only gradually surrendered.

Several key features of the Chinese reform path over the past twenty years have been responsible for growth in the livestock sector. Trade was liberalised and, together with new investment by both state and foreign capital, this led to an increase in manufacturing livestock products of wool, cashmere and leather. Demand for these raw materials expanded as small- and large-scale industries, often with foreign investment, flourished. New technology could be imported in the manufacturing industry. For example, Chinese wool processing capacity increased fivefold in the 1980s (Brown and Longworth 1995). This led to rising producer prices for raw goods and later for manufactured goods exported to international markets. The household responsibility system allowed family livestock farms to sell part of their output at market prices, while compulsory state purchase prices were also raised (Benson and Svanberg 1998). As China's urban population became more prosperous with the general economic growth, demand and prices for meat rose, encouraging livestock producers to keep and sell more animals (Hamann 1999). Thus, Kazak pastoralists in China's Xinjiang province, for example, have seen their incomes double over the past decade, while livestock numbers continue to increase in this province and elsewhere in China's rangelands.

Policies favouring economic growth in China's livestock sector also arise from the state's concern to avoid the political dangers of an impoverished peasantry by improving rural living standards in general (Pomfret 1997). In regions with significant groups of pastoralists from ethnic minorities, there has been an additional concern that herders' incomes should keep pace with the crop-based and urban Han Chinese majorities.

Comparisons with agrarian reform in Mongolia

Mongolia presents a seemingly contradictory case, in which rapid reform has been accompanied by a considerable increase in pastoralism and livestock numbers since the early 1990s. A former Soviet satellite, the independent Mongolian government moved swiftly to decollectivise the state livestock farms, liberalise prices and withdraw state inputs (Collins and Nixson 1993; Pomfret 2000b). This was a pattern repeated in the former Soviet states of Kazakstan and Kyrgyzstan, but where disastrous results had ensued for the livestock sector, as already discussed. However, in Mongolia not only did pastoralism survive these radical transition programmes but the livestock sector has continued to grow throughout the last decade, similarly to China but at an even greater rate (see Table 2.2). In this period, the number of registered herders in Mongolia has more than trebled, from 18 to 50 per cent of the national workforce (Sneath 2000). Making a living by herding livestock has become more economically attractive following agrarian reform. There are two questions: why this occurred in Mongolia but not in the other ex-Soviet controlled nations of Central Asia, and second, how has this been achieved under conditions of rapid rather than gradual reform as in China and southeast Asia.

The first question concerns two separate issues – why did the pastoral economy not initially collapse, and then why did it subsequently grow? It is possible to shed some light on these issues as the functioning of the pastoral economy has been well-documented over the past decade in Mongolia, in contrast to ex-Soviet Central Asia. The reason why pastoralism overcame the shock of rapid reform lies in the particular state management goals of the previous socialist period (Neupert 1999; Sneath 2000). In Mongolia, collectivisation of pastoralists was introduced much later, only during the early 1960s, compared to Central Asia. Thus, the mobile form of husbandry and the social institutions supporting pastoralism were not as severely disrupted as in Soviet Central Asia. Moreover, a crucial difference was in the degree of intensification intended by the central government for the livestock sectors in Mongolia compared to the Central Asian states.

Under Soviet influence, Mongolia was never planned to be a rural factory producing agricultural surpluses to feed and clothe the Soviet Union, as for example, were Uzbekistan and Kazakhstan with their cotton, wheat, meat and wool.

Table 2.2 Increase of livestock in Mongolia and China, 1990–9

Livestock	Mongolia Percent increase 1990–9	China Percent increase 1990–9
Cattle	37	35
Sheep	3	12
Goats	122	44
Horses	41	−13
All the above	35	28

Source: FAO 2000.

24

Certainly, there was some intensification of the traditional Mongolian livestock husbandry practices, mainly in the form of introducing fodder, but this was largely as a by-product of greatly expanded crop farming. The fodder was planned to provide emergency feed supplies during particularly harsh winters (Templar *et al.* 1993). But the pastoral system of production continued to rely mainly on natural pastures even through the winters (Sneath 1999). Thus, Mongolian pastoralism never became as critically dependent on external inputs as was the case in Kazakstan and Kyrgyzstan where livestock farming had been intensified over half a century.

When external inputs ceased to be supplied and the national economy was in crisis, almost immediately after the dissolution of the Soviet Union, Mongolian pastoralists had retained a fairly resilient form of livelihood and could retreat into subsistence following decollectivisation (Pomfret 2000b; Sneath 2000).

The reasons for the recent growth in pastoralism and livestock populations in Mongolia are twofold. First, loss of employment with state organisations led many urban dwellers to take up livestock husbandry, either directly or indirectly, as a source of income (Fernadez-Gimenez 1999; Pomfret 2000b). Such a strategy was possible because of kinship linkages between urban and rural households, and constituted a survival strategy under conditions of economic breakdown in the formal sector. Some households also began to invest in livestock as the economic returns, particularly for high-value cashmere goats, were attractive relative to other savings options.

This is the second reason why livestock numbers have grown: the commercial value of cashmere, which is produced by goats. By far the greatest increase in animals has been in the goat population, which more than doubled during 1990–9 (Table 2.2). In that period there has been major expansion in the commercial exploitation of cashmere, as demand from China has boomed and foreign companies have invested in cashmere processing within Mongolia (McGregor 1996; Mission of Mongolia 2000). The growth of livestock production has been enabled by the policies of free trade and incentives for foreign investors adopted since 1995, with an emphasis on development led by the private sector. As a result, the wool and cashmere processing sector currently contributes over 10 per cent to total industrial production, while the leather industry accounts for a further 15 per cent. Domestic processing of leather and wool declined up to 1998, however (Sneath 2000). Production of sheep wool, camel and goat hair trebled from 1993 to 1997, and their export values increased in value during the same period (Mission of Mongolia 2000).

The overriding factors responsible for the present generally positive condition of the livestock sector has been commercial and export-oriented economic policies. In this respect, Mongolia's national reform programme has resembled that of China. On the other hand, agrarian reform had generally negative effects. Decollectivisation of state livestock farms removed state support from herders, leaving them vulnerable to climate disasters and higher costs of production. Price decontrol raised the cost of living sharply as food and fuel prices rose. Rapid reform initially had similar effects on Mongolian pastoralists as in Kazakstan and Krygyzstan. However, Mongolian pastoralism has withstood the shocks of agrarian and price reform because herders could continue raising animals when state support was first withdrawn, and subsequently had opportunities for selling high-value fibres within an enabling economic and business environment.

Conclusions

This overview has highlighted a number of patterns in how different types of reform have affected the pastoral livestock sectors in former socialist countries of Asia.

Impacts have been conditioned by the pre-existing organisation of agriculture as well as by the goals and speed of state reform programmes. With hindsight, it was a mistake to suddenly undercut the foundation of high input-dependent agricultural systems that were vertically integrated into a closed market. A high degree of dependence upon the state, indicating a lack of self-sufficiency among capital-intensive livestock farms, requires a slower weaning process if new self-reliant systems are to have time to evolve. This is the explicit strategy of the slow reforming countries of Turkmenistan and Uzbekistan. As it was, much of the former infrastructure of the livestock industry in Kazakstan and Kyrgyzstan was destroyed in the reform process before there was time for new forms of production and marketing to emerge. Livestock output has fallen greatly along with employment and incomes in the livestock sector.

A second lesson to emerge from the experiences of the reforming socialist countries is that opening up trade barriers and encouraging investment into agro-processing pays dividends by increasing the value of agricultural products, thus giving incentives for farmers to carry on. The economic growth of China and Mongolia's livestock industries demonstrates the importance of export-led and business-friendly policies.

Arising from this is an appreciation that the immediate disruptions resulting from a rapid pace of reform – the so-called 'shock therapy' – can be mitigated so long as strong demand and attractive prices exist for agricultural products. In those countries with relatively small internal markets and poorly developed manufacturing capabilities, the source of this demand may be external at least until domestic processing can be modernised. But being able to respond to demand from whatever source allows farmers to secure their livelihoods.

Part I

PASTORALISTS AND RANGELANDS OF SOUTHEAST KAZAKSTAN

Farming settlements in the study area are composed of relatively small, nucleated villages, with several hundred to several thousand people. The great majority of residents are now ethnic Kazak. The history and social organisation of the Kazaks is presented in studies by Olcott (1995) and Akiner (1995). Prior to independence, these settlements contained a proportion of families from other ethnic groups, including Russian, German and other ethnic groups of the former USSR. Most of the non-Kazak families have left the villages in the study area, while some ethnic Kazak families have moved into these villages from neighbouring countries such as China and Mongolia.

Families have from four up to ten or more children. Younger children are expected to assist with agricultural and livestock-management tasks, while older children who have left school often attend institutes of higher education in the main city, Almaty. Married sons and their wives usually continue to reside with the son's parents until some years after the birth of their children and in some cases until the death of the parents. Bride price is given, usually in the form of livestock, by a groom's parents to the bride's family. Married sons typically have a joint share in their parent's livestock and crop land, and manage these together with their parents and other siblings. When daughters marry, they settle in their husband's home and contribute their labour to their husband's family.

There are strong traditions, still upheld, of mutual assistance between kin. This assistance involves exchanges of goods, services and money, extending within and between villages, as well as between rural and urban kin. The operation and function of these linkages are described by Werner (1998) for southeast Kazakhstan.

Tasks of livestock management are assigned according to age and gender. Adult men or older boys herd and take animals to water, usually on horseback. Women are responsible for milking animals, processing milk into a number of different products, and handling wool and fibre. Children as well as women and men provide

27

winter fodder to the animals in the barns. Men, and sometimes women, take the family livestock to urban markets.

The settlements in the study area each belonged to a state farm (*sovkhoz*), administered from a central village which contained the majority of the population attached to the particular farm. The *sovkhozes* were dissolved in 1995/6. The central villages housed the farm administration offices, which supervised most activities in the community. These villages were service centres, with primary and secondary schools, a state grocery shop, a cultural centre, medical clinic (in the case of larger centres, small hospital), electricity sub-station and connections to the national telephone grid. Houses were provided, of modern construction with central heating by solid fuel, but no indoor sanitation. Water was piped to street outlets. As part of the farm services, each centre had a technical department that maintained agricultural equipment and vehicles, and a veterinary section. Larger state farms also had outlying smaller villages, or brigades, in which resident workers carried out specialized tasks such as harvesting hay or growing crops.

Most of the former *sovkhozes* in the study area grew rain-fed and irrigated crops in addition to raising livestock. In the areas of the semi-steppe and foothills, with between 250 and 450 mm annual precipitation, wheat and barley can be grown as well as fodder crops of maize, alfalfa and clover. In drier areas of the northern part of the study area, pastures were improved by sowing *Agropyron* and other wild species, while a limited amount of human and animal food crops were grown under irrigation.

Every working family caring for the state farm livestock had a house in the central village, from which children could attend school, and to which the shepherding families could return in between managing their assigned animals at disant pasture areas. During the winters, shepherding families had to stay in small houses up to 50 km or more distance from the central village. During the summer some shepherding families would again move from the central village to different pastures where they lived in yurts – felted tents. Transport to and from the distant pastures was provided by the state farms. In spring and autumn, shepherd men could sometimes stay at the village home if the pastures were not too distant. Older people and school-age children spent most of their time in the villages, while husbands, wives and younger children moved continually between the winter houses, the main village home and the summer camp.

Approximately 80 per cent of Kazakstan's agricultural land is pastures, most of which are semi-arid with annual precipitation of less than 300 mm. The total area of pastures is 189 million ha. By the end of the Soviet era, a further 5 per cent of agricultural land was devoted to fodder crops. Livestock comprised 60 per cent of the annual agricultural output. There were between 7,000 and 8,000 state farms, and the state bought up to three-quarters of all production. The remainder was consumed on farm as food or feed, used in barter trade or sold on the black market (World Bank 1993). The pastures remain the principal landed resource of the country.

The study area

The research was carried out along a transect of some 300 km length which includes some of the major ecological zones grazed by livestock in Kazakstan. The transect covers each of the seasonal pasture areas that were used annually by the former state collective livestock farms (*sovkhozes*). The transect is located in the western portion of Almaty *oblast* (administrative region), mostly within Dhambul *raion* (administrative district), part of the Lake Balkhash basin. The transect follows a northwest–southeast gradient beginning in the sand dune desert south of Lake Balkhash, through semi-desert and steppe, and ending in the high altitude meadows of the Ala Tau mountains. There is a precipitation gradient, with less than 200 mm in the north increasing to nearly 700 mm in the southern mountains.

Chapter 3 contains details of the main plant communities in the study area while climate information is given in Chapter 4. Details of livestock in the study area are given in Chapter 5.

3

TRANSHUMANT ECOSYSTEMS

Fluctuations in seasonal pasture productivity[1]

Ilya I. Alimaev

Introduction

Kazak pastoralists traditionally followed a transhumant or nomadic way of life. They made maximum use of natural pastures in different ecological zone areas at each season. This system was reshaped during the Soviet period, and is now considerably disrupted. Yet extensive livestock production based on seasonal movement remains the most efficient way of exploiting the natural resources of Kazak rangelands, as dependence on artificial livestock feed sources is minimised. Thus the prime purpose of new management systems should be to obtain maximum production from pastures with the minimum amount of stall-fed fodder.

The first purpose of the study was to understand the ecological variability that underpins the system of livestock movement practised before and during the Soviet period. Decollectivisation of state livestock farms has significantly changed the way pastures are used as well as removing the state provision of supplementary fodder. There has been a major decrease in the number of animals and a reduction in seasonal livestock mobility after 1994/5, as described in Chapter 5. One of the consequences of these changes is that emerging private livestock management systems are not able to take the most advantage from natural variability in the environment.

The second purpose was to detect any measurable impacts on pasture productivity as a result of changes in livestock numbers and movement. It was hypothesised that changes in the density and location of animals in each season would have different impacts on each ecozone. It is found that the decrease in livestock populations may be leading to a restoration stage on pastures in certain

---

1 This chapter is partly based on a paper entitled 'Mobility and the market: economic and environmental impacts of privatisation on pastoralists in Central Asia' (co-author, C. Kerven), presented at the conference on 'Strategic considerations on the development of Central Asia', Council for Development of Central Asia/Chinese Academy of Sciences, Urumchi, 13–18 September 1998.

ecozones which were much more heavily used in the recent past. These findings from field measurements are comparable to those described in Chapter 4, based on satellite interpretation. However, greater concentration of animals around settlements and water, stemming from a loss of mobility, is having localised damaging effects on pasture productivity.

Reorganisation of transhumance in the twentieth century

Nomadic and transhumant pastoralists were able to exploit the natural resources of Central Asia's steppes, deserts and mountains for some three thousand years (Asanov et al. 1992b; Zhambakin 1995). Exploitation was based on long-distance movements from 600 up to 2,000 kms. Winters were spent in grazing areas less subject to the severe cold and winds, where some forage could be obtained. Such places included sandy dune deserts, sheltered south-facing mountain valleys and areas of shrubs and reeds next to rivers and lakes. In spring, animals were moved to graze on nutritious ephemeral plants of the semi-desert, while summer was spent grazing the grasses of the northern steppes or in the high meadows of mountains in the south and east. Autumn involved a return towards the winter grazing sites, grazing shrubs through the semi-desert along the route. Kazak pasture scientists emphasise that the sandy deserts were a key resource for pastoralists as in winter the micro-topography and shrubby vegetation provided both forage for animals and fuel for humans (Asanov et al. 1992b).

The movement patterns of Kazak pastoralists were compressed during the first period of Russian administration beginning in the 1850s. In particular, areas of pasture land were alienated by Russian settlers, and some settlement of the nomads took place. In the first part of the twentieth century, the main impact of the October Revolution on pastoral mobility was the return of some pasture land to the Kazaks, and state assistance in establishing agriculture (Zhambakin 1995). Livestock populations increased during this time. The first collective farms, in which members retained their own livestock, were established in this period.

In 1928, Stalin decreed that livestock should be expropriated from wealthy Kazaks and, in 1929, that ordinary Kazak pastoralists should be collected and settled into state farms in one year (Zhambakin 1995). Kazaks responded by slaughtering their animals or by escaping with their animals to neighbouring countries. Many of the remaining animals which were collectivised died due to a lack of winter feed. The result was a huge loss of livestock in the country and great famine among the population.

After the catastrophic loss of livestock in the early 1930s, numbers increased steadily during and after the Second World War, particularly during the 1950s. The state re-introduced a system of long-distance seasonal migration. Land was added to the collective farms. Land on the plains was to be used for spring and autumn pasturing, and land in the sandy desert used for winter pasturing. State farm animals were again taken to these distant pastures each season. By the end of the 1940s half of all state animals were managed by the long-distance migration system.

For some time, there was a period of prosperity for the pastoral livestock sector. The system of seasonal use of distant pastures provided a cheap, nutritious and high-energy feed resource, which resulted in maximum population growth and high-quality outputs with a low labour input.

Intensification and fodder production from the
1960s to the 1980s

In 1965, the state issued a plan to establish some fifty million sheep and goats in the country. To achieve this goal, some 150 new state sheep farms were organised in the semi-desert and desert zones on land hitherto only lightly used by season. Permanent farm boundaries obliterated nomadic seasonal migration routes and livestock were pastured for at least three and sometimes four seasons on the same areas.

The plans to increase livestock numbers required additional feed resources. These were provided by scientific research which increased fodder production and developed sown pastures. The low natural productivity of semi-arid pastures was enhanced by planting improved varieties of nutritious and hardy forage species, *Agropyron*, *Elymus*, *Kochia*, *Salsola*, *Eurotia* and others, over a total of six million ha. Research also resulted in additional forage production through use of fertilizers on mountainous and flood-plain areas, use of maize as silage and cultivation of fodder plants such as clover and barley, often under irrigation. The total increase in fodder production over thirty years to 1983 is shown in Figure 3.1.

A form of industrialised semi-nomadism termed 'Mechanised livestock-stations' was thus implemented by the state, in which the inherently low carrying capacity of the pastures was conquered by partly substituting traditional seasonal mobility with modern scientific techniques.

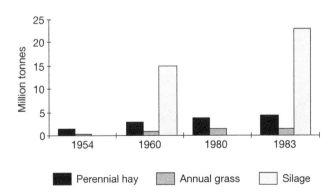

Figure 3.1 Increase of fodder production in the Soviet period.

Nationally, at the end of the Soviet period livestock obtained half of their feed requirements by foraging on pastures. The ratio from foraging in south and southeast Kazakstan was considerably higher, from 75 to 100 per cent, due to a warmer climate (Nordblom *et al.* 1996).

The remaining feed requirements were cut and stall-fed to animals from two main sources – natural haylands and cultivated lands – with the latter consisting of irrigated and rain-fed land. Prior to independence, natural hayland produced between 14 to 15 million tonnes of hay on 5.8 million hectares mainly in flood plains. Haylands had high nutritional qualities (see Table 3.1). In addition, fodder crops of alfalfa, maize and others were grown on 10.4 million hectares of arable land, contributing 30 per cent of total livestock feed requirements. However, the productivity of fodder crops on irrigated land was low, at 10–25 fodder units/ha. Relative to cost of production, this was a less efficient source of animal feed than natural hay (see Appendix I on page 257 for the definition of fodder units). Values of different types of fodder are shown in Table 3.1.

Since decollectivisation, new private livestock farmers can usually not afford to produce high quality fodder. Although harvested hay is now probably the most economical form of providing winter feed, lack of agricultural equipment prevents many farmers from harvesting at the optimal time. This means a loss of feed value. For example, in the case of *Agropyron* there is a drop from 5.3 per cent protein and 10.5 per cent cellulose at heading phase to 1.7 per cent protein and 15.9 per cent cellulose when seeded out.

The main limiting factor on livestock production in Kazakstan has always been the extreme winters. In the study area, mean January temperature is $-12\,^{\circ}$C, with an average snow depth of 16 cm. Supplementary fodder is required at various times and in varying degrees over winter, to maintain animals when pastures are inaccessible for grazing due to snow or ice cover. With the protective mechanisms of the state no longer available in the form of fodder provision and social security, the majority of producers now face high risks as well as costs. Following privatisation

Table 3.1 Nutritive value of some main types of hay produced by state farms in western Almaty *oblast* (per kg dry matter)

Parameters	Mountain: grass and forbs	Steppe: grass and Artemisia	Sown Agropyron sibiricum	Sown alfalfa Medicago sativa
Metabolic energy in MJouls	8.3	7.9	6.7	8.6
Fodder units	0.68	0.48	0.52	0.54
Total protein, % dry matter	8.9%	7.9%	8.9%	17.6%
Digestible protein, % dry matter	6.2%	4.5%	6.6%	12.1%

of the livestock industry in the 1990s, the lack of winter fodder and its low quality in general has again become the major limiting factor for livestock husbandry.

Methods of study

A transect of approximately 300 km length and 100 km width was defined to include five distinct ecological zones. Each zone was formerly grazed in one season as part of a regional transhumant ecosystem in western Almaty *oblast*, up to the end of the Soviet period.

The methodological approach was to compare past and current indices of vegetation productivity and fodder value by season in each ecozone. One, or in some cases two, typical sites were selected to represent each of the five ecozones. Plant samples were taken at each site seven times, in the spring, summer, autumn and winter of 1998 and the spring, summer and autumn of 1999. The samples were then analysed in the laboratory of the Kazak Scientific Institute of Pasture and Fodder to determine total biomass, species composition, feed and chemical values. Methods of plant collection and assessment are given in Appendix II (page 257).

In mid 1998, a second set of three sample points was added, to compare the effects of grazing on vegetation near and at some distance from a water point. This second sample was to assess the effects of greater animal concentration around the few remaining water points after the collapse of infrastructure accompanying decollectivisation. Measurements of pasture plant productivity were taken at these three sample points in autumn 1998, and spring, summer and autumn of 1999.

Two years is an insufficient period to record long-term vegetation changes associated with the scale of change in livestock management systems found in the study area. However, some general trends from the results may be observed. The work carried out will also provide a basis for long-term monitoring in the future.

Ecological zones along the transect

The transect follows a northwest–southeast precipitation gradient beginning in north in the sand dune desert around Lake Balkhash, through semi-desert and steppe, and ending in the south in high altitude meadows of the Ala Tau mountains. The agro-ecological characteristics of each zone is described below. Botanical assessments in each zone by season are shown in Table 3.2 (see also Chapter 4).

Sand desert

The northern end of the transect comprises the Taukum sands, with a mixture of sand forms including hillocks, small and large dunes. Vegetation is a mixture of semi-shrubs (mainly *Haloxylon* spp.), with ephemeral grasses and *Artemisia* species in the depressions. Annual precipitation is between 125 and 135 mm of which up to 80 per cent falls in spring. This zone was formerly used as winter pasture due to the accessibility of vegetation above the snow level on exposed ridges and south-facing dune slopes. In spring the nutritive value of the pastures is very high.

Semi-desert plain (north)

Following the transect south, the next zone is a semi-desert plain, with annual precipitation of 225–50 mm, dominated by *Artemisia terrae-albae*, accompanied by ephemeral grasses and dwarf semi-shrubs. The vegetation reaches its peak in June, and was heavily grazed in late spring during the latter Soviet period.

Semi-desert (south)

Another semi-desert zone lies to the south, on more rolling countryside with differing vegetation associations, mainly *Artemisia* with *Kochia, Salsola, Ceratocarpus* and various ephemeroids. Annual precipitation averages 200–50 mm mostly in spring and autumn. This zone provides luxuriant spring vegetation which dies back in summer due to heat, and becomes more productive in autumn when temperatures are cooler and after autumn rainfall.

Dry steppe

A band of dry steppe with annual precipitation of 300–400 mm stretches beneath the foothills and mountains of the Ala Tau. This zone is relatively settled farmland, on which barley, wheat and fodder crops are planted, providing the main source of cultivated fodder for livestock in the whole region.

Sub-alpine meadows

Mountain meadows form the southernmost portion of the transect, at altitudes of 1900 to 2200 metres above sea level. Annual precipitation is approximately 750 mm falling mainly in spring and summer. The humus-rich loamy soils contain a mixture of grasses and legumes, which serve as perfect pastures for the short summer period.

Ecology along the transect

Some of the critical physical factors that influence pasture production along the study transect are summarised in Figure 3.2. Annual precipitation and the humus content of the soil decreases with declining elevation. Combined with differences in groundwater availability, these variable physical factors have produced a wide variety of plant communities within a relatively small geographical area, as shown in Table 3.2.

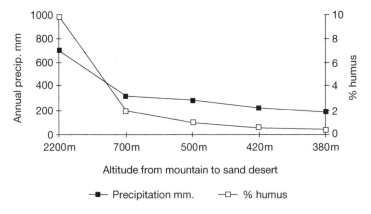

Figure 3.2 Climate and physical features along study transect.

Seasonal variability of pasture productivity

Species of fodder plants have different life cycles according to the ecological zones in which they are located, as shown in Tables 3.3 and 3.4. The combined effects of seasonal variability are shown in Figures 3.3 and 3.4. Ephemeral and ephemeroid plants are mainly found in the drier areas of the rangelands – the desert and semi-desert. Their growth period in spring from beginning of growth to drying out is 50–60 days, but in some humid years this extends up to 75–80 days. Ephemeral plants with short growth periods provide livestock with nutritive fodder in spring and facilitate their speedy fattening after the winter. In humid years, which occur every three to four years along the transect, spring productivity of such pastures is 150–80 kg/ha of dry mass; 1998 was one of these humid years.

Vegetative growth in the mountains starts in spring as a rule, after the snow melts, and starts first on southern slopes and fifteen or twenty days later on northern slopes. The mountain pastures produce very high levels of biomass in the summer, due to the higher rainfall and richer soils. Summer output from these pastures was over 2,500 kg/ha dry mass in 1998. Grazing these pastures give livestock an opportunity to gain weight and fatten in preparation for the mating season in autumn, and for the lean period in winter (see Chapter 6).

In the autumn, late-flowering steppe and semi-desert pasture species (*Artemisia* spp. and others) provide another peak of biomass for grazing, as temperatures fall and some rainfall occurs. Autumn output in 1998 from these pastures was over 1000 kg/ha, higher than in the preceding spring and the same as summer 1999.

In winter, tall shrubby plants (*Haloxylon* spp.) in the sand desert can provide browse for livestock, as total output and protein content is relatively stable in winter despite low annual precipitation. Thus winter output in 1998 was about 500 kg/ha, equal to spring output and only slightly lower than summer and autumn output (see also Chapter 4).

Table 3.2 Vegetation associations and sample sites across the transect, 1998/9

Ecozone	Position	Plant association	Sample dates 1998/9	Main species in sample harvest	Condition
Sands (1)	N44° 46.15' E75° 14.02'	Woody – ephemeral, shrub – forbs	Winter (Feb)	*Artemisia Eurotia Kochia*	Snow cover 8–15 cm South slopes of dunes have thawed patches Snow melted around big shrubs
			Spring (May)	*Haloxylon persicum Kochia prostrata* and *Ceratoides* = 79 % *Carex physodes* = 21%	Coverage 28% State of shrubs good Ephemeral vegetation dominated by *Carex* All plants in good state
			Summer (August)	*Kochia prostrata* and *Ceratoides* and *Artemisia* Total = 93%	Coverage 14%
			Autumn (October)	*Kochia, Ceratoides* and *Artemisia*; one third each.	Coverage 17%
Sands (2)	N44° 29.99' E75° 26.45'	*Artemisia – Agropyron – Ceratoides – Calligonum*	Winter	*Agropyron, Kochia, Eurotia, Artemisia*	Snow 8–13 cm deep. Green leaves of *Agropyron* 5–12 cm long
			Spring	*Agropyron sibiricum* = 90%	Coverage 31%
			Summer	*Agropyron sibiricum* = 54%	Coverage 24%
			Autumn	*Agropyron sibiricum* = 71% Semi-shrub shoots = 29%	Coverage 21%

Plain desert	N44° 09.34' E75° 51.74'	*Artemisia* – ephemeral	Winter	*Artmesia* and *Ceratocarpus*	Light snow cover
			Spring	*Artemisia terrae-albae* = 68% *Eremopyrum orientalis* *Carex* *Tulipa* *Papaver* *Kochia* *Ceratoides*	Coverage 77% *Artemisia* growing well. Other plants abundant and completely foliated
			Summer	Not recorded, only total harvest *Artemisia terrae-albae*	
			Autumn	*Kochia* *Ceratoides*	Growth of *Artemisia* is poor. A few ephemeral plants.
Semi-desert (1)	N44° 01.33' E75° 31.82'	*Artemisia* – saltwort – ephemeral	Winter	*Artemisia, Camphorosma, Kochia*	Northern slopes had snow patches
			Spring	*Artemisia sublessingiana* and *Artemisia terrae-albae* = 70% Ephemeral = 30% *Isatis emargnata*	Coverage 79% Growth of emphemous plants favoured by combination of warmth and humidity
			Summer	Not recorded, only total harvest	Vegetation damaged by locusts. Photosynthetic activity ceased, as all leaves consumed
			Autumn	*Artemisia sublessingiana* near total, small amount of *Ceratocarpus*	Coverage 53% *Artemisia* growing well

continued . . .

Table 3.2 continued

Ecozone	Position	Plant association	Sample dates 1998/9	Main species in sample harvest	Condition
Semi-desert (2)	N43° 48.43' E75° 32.47'	*Artemisia* – grass – ephemeral	Winter	*Artemisia, Ceratocarpus, Kochia*	Snow cover
			Spring	*Artemisia sublessingiana* = 61% *Festuca* sp. *Eremopyrum orientalis* *Poa bulbosa* *Stipa cappilata*	Coverage 74%
			Summer	Not recorded, only total harvest	
			Autumn	*Artemisia sublessingiana* = 100%	Coverage 58% Growth of *Artemisia* is poor Ephermous plants are absent
Steppe	N43° 20.67' E75° 45.12'	Feather grass – *Artemisia*	Winter	*Stipa, Artemisia* and remains of ephemeral plants	Snow cover
			Spring	*Artemisia sublessingiana* = 52% *Stipa cappilata* *Bromus* (perennial grasses = 48%)	Coverage 89% Plants in good state
			Summer	*Stipa, Festuca and Bromus* Total = 75% *Artemisia sublessingiana* = 25%	Coverage 76%

Mountain (2000 m above sea level)	N43° 05.76' E76° 28.75'	Grass – forbs	Autumn	*Artemisia sublessingiana* = 73% *Stipa and Festuca* = 26%	Coverage 66%
			Winter	*Bromus, Poa, Dactylis, Artemisia, Stipa*	Snow depth of 33–57 cm on flat ground. Southern slopes have no snow cover in places
			Spring	*Agropyron repens* = 100%	Coverage 95% Grasses at tillering stage
			Summer	*Dactylis glomerata* = 64% *Bromus inermis* = 23% *Agropyron repens* = 13%	Coverage 100%
			Autumn	*Agropyron repens* = 100%	General appearance yellow, vegetation growth completed

Table 3.3 Dates of phenological phases of dominant pasture plants along the study
transect

Plant species	Beginning of growth–budding	Budding–flowering	Flowering–fruiting	Fruiting
Artemisia terrae-albae	6.04–21.06	21.06–4.09	4.09–27.09	27.09
Artemisia sublessingiana	18.04–22.06	22.06–13.08	13.08–15.09	15.09
Agropyron sibiricum	3.04–29.05	29.05–15.06	15.06–2.07	2.07
Ceratoides papposa (formerly *Eurotia ceratoides*)	14.04–3.07	3.07–3.08	3.08–19.09	19.09
Ceratocarpus arenarius	31.03–1.05	1.05–12.05	12.05–28.06	28.06
Carex physodes	26.03–28.04	28.04–3.05	3.05–17.05	17.05
Poa bulbosa	29.02–20.04	20.04–6.05	6.05–21.05	21.05
Agropyron repens	8.04–6.06	6.06–19.06	19.06–12.07	12.07
Stipa cappilata	13.04–19.05	19.05–5.06	5.06–17.06	17.06
Kochia prostrata	4.04–26.06	26.06–22.07	22.07–3.09	3.09
Bromus tectorum	22.03–25.04	25.04–15.05	15.05–22.05	22.05

Table 3.4 Dates of phenological phases of fodder plants of mountain pastures in study
transect

Plant species	Beginning of growth	Beginning of tillering of grasses	Beginning of flowering
Trifolium repens	13.04	–	24.05
Agropyron repens	28.04	30.07	11.08
Alopecurus arundinaceus	19.04	6.06	22.06
Poa pratensis	19.04	7.06	27.07
Koeleria cristata	22.04	15.06	20.07

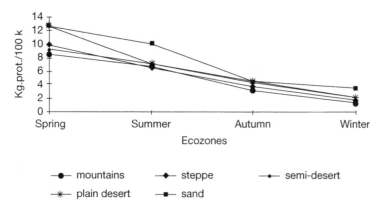

Figure 3.3 Protein availability by season in each ecozone, 1998/9.

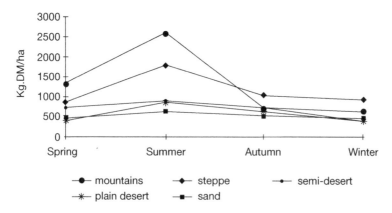

Figure 3.4 Pasture yield by season in each ecozone, 1998/9.

Ecology, land use and livestock mobility

The ecological variation that provides the logic for livestock migration is apparent
from Figures 3.3 and 3.4. Desert and semi-desert pastures have a sudden, short
and high protein level in spring. Protein levels and total output in the steppe are
at their maximum in summer. Mountain pastures are extremely productive over
the summer, but their protein levels do not match those of drier areas in the spring,
and they cannot be used over winter or spring due to snow cover. The semi-desert
and desert achieve a second lower peak of production and protein output in the
autumn. The carrying capacity of different pasture types thus varies by season.
Maximum efficient use of the total pasture resource can be achieved by integrating
seasonal variability through movement. No single pasture type can provide the

annual carrying capacity equivalent of combining use of each pasture type in sequence. Livestock migration partially levels out the imbalance between the quantity versus the quality of different forage resources. As shown in Figures 3.5, 3.6 and 3.7, the total biomass output of the dry areas is inversely correlated to their nutritional value in the same way as the high level of output from wetter mountain pastures is associated with a lower protein feed value.

The differences in the quantity and quality of output among these plant communities at different seasons has provided the foundation for mobile forms of livestock production. Within this area prior to decollectivisation in 1995, most state farm livestock, mainly sheep, followed a four-season grazing migration system. Animals were managed in winter by shepherds in remote quarters located in the sandy dune areas of the northern desert. Here the animals received supplementary feed in barns and grazed in the open on woody shrubs whenever weather permitted. In spring, by March, animals were moved further south to the semi-desert or the steppe, to graze mainly on ephemerals, where lambing and shearing took place. By June, many of the animals were taken for the summer up to mountain meadows where shepherd families lived in yurts (felt-covered tents). In autumn, shepherds

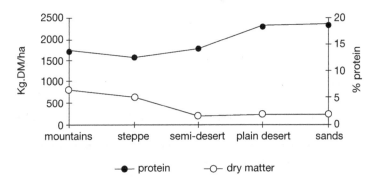

Figure 3.5 Relationship between pasture protein and output, spring 1999.

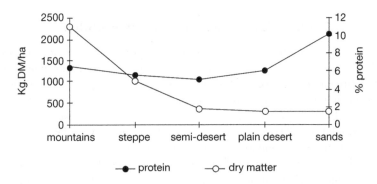

Figure 3.6 Relationship between pasture protein and output, summer 1999.

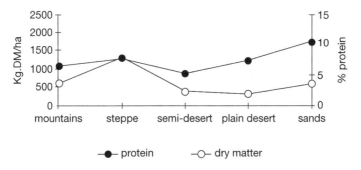

Figure 3.7 Relationship between pasture protein and output, autumn 1999.

then took the animals to different areas in the semi-desert or steppe zones, where breeding took place and animals were culled for slaughter. By the time the first heavy snow fell, animals were taken back to the northern winter pastures. The total length of the migration route was between 200 to 300 km in the study area.

Effect of livestock concentrations around water points

The breakup of the large state farms has led to a great decrease in the numbers of animals and the remaining animals are now moved much less than in the past. Some newly privatised shepherds have settled all year round outside villages on former state farm seasonal sites where water, housing and animal shelters are available (see Chapter 5). The few animals still under the management of ex-state farms – now cooperatives – are also moved much less due to diseconomies of scale in moving small flocks (see also Chapter 7). Thus animals are often now concentrated around water points without being moved away by season as in the past. While the decreasing number of animals is relieving grazing pressure on distant pastures, the process of concentration is building up in localised sites.

To begin to measure the effect of animal stocking on pasture output, seasonal measurements were taken at three sites over a year, at increasing distance from water points. The results for two sites are given in Table 3.5, and Figures 3.8 and 3.9. These show that, while pasture output is depressed immediately in the vicinity of water points, in all but the autumn period output is similar at both 2 km and 10 km from water points. However, in autumn 1998 there was higher rainfall, and pastures furthest from the water point were able to respond with higher output. These results suggest that localised grazing pressure is having an effect on pasture output as sheep are grazed less than 10 km away from water points. While these results are only for a one-year period, they are consistent across two sites separated by about 50 km.

Table 3.5 Effect of livestock concentrations at three pasture sites, 1998/9

Ecozone and plant association	Position	Plant count and main species Autumn 1998–Autumn 1999	Sample dates 1998/9	Harvest dry matter kg/ha	Livestock grazing pressure and other site characteristics
Plain desert Site 1: 500 m from water point	N44° 20.36' E75° 29.92'	Single plants of *Artemisia terrae-albae* and *Ceratocarpus*	Autumn Spring Summer Autumn	147 70 110 180	An artesian well, two families settled for past 19 years. Now grazed all year round by herd of 60 horses, herd of 70 camels and private flock of several hundred sheep. Site used to be only used as winter grazing area for a state farm, which sent all animals elsewhere in other seasons. The state farm used to send 300 horses and many hundreds of sheep to this site in winter. According to shepherds, there are now more pasture plants as fewer animals are grazing now. Only weak animals received supplementary feed in winter 1998/9
Site 2: 2 km from water point	N44° 20.48' E75° 29.40'	*Artemisia terrae-albae* 6 plants sq/m *Ceratocarpus* 6 plants sq/m *Kochia* 1–2 plants sq/m	Autumn Spring Summer Autumn	352 240 290 270	
Semi-desert Site 1: 500 m from water point	N44° 01.43' E75° 14.91'	*Artemisia sublessingiana* 2 plants sq/m *Poa bulbosa* and *Kochia* 1 plant sq/m	Autumn Spring Summer Autumn	130 153 170 230	Artesian well at foot of mountain range. Area was mainly used in past as winter grazing for state farm livestock from the mountains. Pastures on the plain now used by a private shepherd family settled two years ago with 700 sheep and 70 cattle. Sheep now graze on the plain
Site 2: 2 km from water point	N44° 02.38' E75° 15.53'	*Artemisia sublessingiana* 9 plants sq/m	Autumn Spring	346 230	

continued . . .

Site	Coordinates	Vegetation	Season		Notes
		Autumn 1999: Artemisia, Kochia, Poa	Summer Autumn	380 430	three seasons out of four (except winter). Plains have many plants with good energy for animals. In winter mainly *Kochia* and *Artemisia*, in spring *Trigonella*, in autumn *Ceratocarpus*. In winter all animals given supplementary purchased feed of *Agropyron* and grain

Semi-desert

Site	Coordinates	Vegetation	Season		Notes
Site 1: 500 m from water point	N43° 49.36' E75° 52.53'	*Artemisia sublessingiana* 4–6 plants sq/m	Autumn Spring Summer Autumn	122 302 250 230	Spring water and wells. Site used by cooperative (former state farm) as autumn grazing area, but now with many fewer sheep than in the past. One cooperative has two flocks of 400 sheep each grazing over autumn. Two other cooperatives also use the area now for grazing. In recent years, one flock of 600 sheep stayed in same area for one and half years, the other flock of 600 sheep used the area only in autumn.
Site 2: 2 km from water point	N43° 49.59' E75° 52.48'	*Artemisia terrae-albae* and *A. sublessingiana* 9–16 plants sq/m Autumn 1999: Ephemera consumed by locusts in summer. *Ceratocarpus, Kochia, Eurotia* all present	Autumn Spring Summer Autumn	420 340 360 400	Until two years ago, shepherds for these cooperatives grazed animal around several other water points over autumn, but these water points now not maintained, so all cooperative flocks must use single water point. In the 1960s large tracts of the plain were planted with *Agropyron* and barley for harvesting in autumn as winter feed. This is no longer done. In autumn, animals mainly graze *Kochia* and *Artemisia*

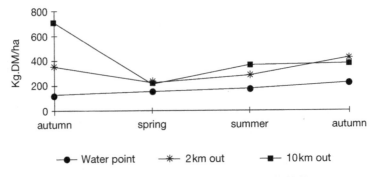

Figure 3.8 Effect of water points on semi-desert pasture output, 1998/9.

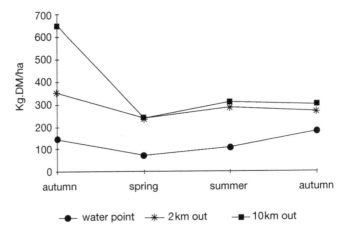

Figure 3.9 Effect of water points on desert plains pasture output, 1998/9.

Long-term changes in pasture productivity

The lower density of animal numbers since the mid 1990s may have begun to have a positive effect on pasture productivity. This is suggested by the results in Figures 3.10, 3.11 and 3.12, which show that average long-term pasture output in all ecozones was considerably lower than that recorded in the two study period years of 1998 and 1999. These data must be interpreted with great caution, however. First, sample sites and dates of sampling were not the same between the data sets. Second, a two-year study period does not provide a sufficient period within which to measure long-term pasture recovery, although animal densities had been lowered for several years prior to the study. Third, a data set of two years is inadequate to even out the inter-annual variability of precipitation, temperature and other factors which will affect pasture output. Nevertheless, local observations and comments by users – the pastoralists – all point to a recovery in some ecozones, due to the vastly fewer number of animals since the mid 1990s.

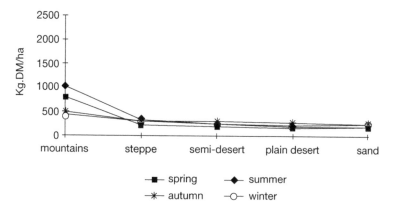

Figure 3.10 Long-term pasture productivity along transect, 1954–72.

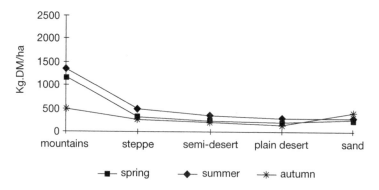

Figure 3.11 Long-term pasture productivity along transect, 1974–86.

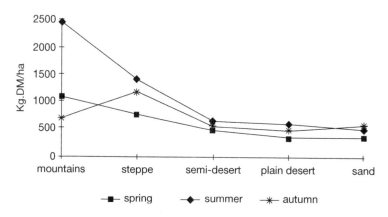

Figure 3.12 Current pasture productivity at study transect sites, 1998 and 1999.

The ability of semi-desert pasture plants to recover from grazing pressure has been measured in the past by the Kazak Scientific Institute of Pasture and Fodder. In the plain semi-desert with 200–25 mm annual precipitation, *Artemisia terrae-albae* was restored under reserve conditions over a period of ten years and in fifteen years under moderate grazing. Under higher precipitation conditions of 300–50 mm in the dry steppe, restoration of *Artemisia sublessingiana* and *A. songarica* Schrenk was achieved within five to seven years under nil or moderate grazing. In contrast, earlier studies in mountain meadow pastures in a grazing reserve indicate that reduction of grazing does not increase productivity of the yield.

These results can be compared to those from analyses of climatic and satellite data for the same study transect (Chapter 4). Those results indicate that sites in the steppe and plain semi-desert are experiencing a recovery in vegetation in the 1990s. These areas were subject to heavy grazing during the latter part of the Soviet period, but were rapidly released from grazing pressure by the sharp decline in livestock populations. Figure 3.12. also shows a rise in pasture output in 1998 and 1999 in these two ecozones, with a particularly steep rise for the steppe.

The mountain site in our assessments also showed a major increase in pasture output compared to the previous forty years, but the same site could not be included in the long-term analysis presented in Chapter 4. Nevertheless, findings in Chapter 4 suggest that the mountain pastures site, as well as sandy desert sites, were not subject to serious degradation in the Soviet period and would therefore not be expected to show signs of recovery in the present period.

Conclusions

The area represented by the study transect is a natural complex that fully corresponds to the requirements of extensive mobile livestock husbandry. Availability of different types of seasonal pastures allows livestock to be grazed almost all year round. In addition, soil and climatic conditions in the area allow producers to harvest and store fodder for supplemental winter stall-feeding. Given the natural environment and current constraints, a system of mobile livestock husbandry would still be economically and environmentally optimal. However, major upheavals in the organisation of livestock farming, resulting from the decollectivisation of state farms and rise of private farming, have broken apart the system of transhumance previously pursued. Most fodder plants along the transect are distinguished by high palatability and high nutrient content. Current management of pastures does not allow sufficient advantage to be made of these resources.

The pastures along the study transect were intensively exploited up to the early 1990s and some zones have been strongly degraded. The subsequent decrease in animal numbers has caused the beginning of a restoration stage in certain areas, notably the steppe and plain semi-desert pasture zones. This suggests that at least some types of pasture have the ability to recover, and is a positive trend deserving appropriate policy measures and technical support. However, as the evidence for recovery is still very slight, further assessments are required to confirm this trend.

An inverse relationship has been shown between pasture productivity and the nutritive value of pastures for livestock; pasture productivity declines in more arid environments, while nutritive values increases. This finding underscores the contribution of low-volume but high-quality desert pastures to a regional livestock grazing system.

At the beginning of the present reform period, large areas of pasture land in Kazakstan were degraded, in part due to high numbers of animals which had resulted in an unbalance between forage resources and animal populations. Some types of pasture had lost their potential efficiency while the quality of their output was also reduced. This situation is now changing but new challenges are presented. These are to promote systems of livestock management which do not again impair the pastures and which lower costs of production for new private farmers.

4

COLLAPSE OF THE KAZAKSTAN LIVESTOCK SECTOR

A catastrophic convergence of ecological degradation, economic transition and climatic change

Jim Ellis and Re-Yang Lee

Introduction

The state-subsidised livestock systems in Kazakstan as in other Central Asian regions successfully met the meat and wool needs of the Soviet Union. However in the mid-1960s and 1970s the central planners in Moscow perceived the need both for increased production from the livestock sector and improved living standards for local herders. The resulting push for yet more intensive production may have instigated a long, slow decline in pasture productivity and ecosystem sustainability. It is likely that unsustainable land use practices and a significant shift in climate patterns in the 1990s also contributed to the demise of the livestock system.

It is shown here that rangeland degradation and increasing climatic variability appears to have been instrumental in the collapse of the livestock sector in the mid-1990s. The economic transition then suspended the flows of fodder and other inputs, leaving the pastoralists in severe economic difficulties, and with a seriously depleted environmental resource base.

Today, livestock and rangeland production systems in Kazakstan and elsewhere in Central Asia are undergoing reform and reformulation, following independence from the USSR. In the current period of uncertainty and experimentation, a variety of new, spontaneously developed herding strategies and range use patterns are emerging (Chapters 5, 6, 7 and 8). Eventually, these will coalesce into new livestock-based land use systems, the sustainablilty and economic performance of which will depend, in part, upon how well livestock management and land use fit the inherent dynamics and production patterns of the natural ecosystem described in Chapter 3.

In this chapter, we explore the recent (1982–98) dynamics of rangelands in the study area covering Balkash basin, southeastern Kazakstan, through analysis of long-term climate data and remote-sensed imagery. The purposes of the analysis are: 1) to describe the structure and dynamics of this rangeland ecosystem, and to relate this to historical and recent land use patterns in the Balkash basin; and 2) to evaluate the evidence that the ecosystem was indeed overexploited and degenerating, or alternately, that pastures were in reasonable condition in the mid-1990s, at the time of the livestock sector decline.

The Balkash basin ecosystem

The Lake Balkash basin is in southeastern Kazakstan, between 43° and 47° north latitude. The basin is a dramatic environment, combining a harsh arid to vary arid, cold-winter, warm-summer climate, with substantial landscape diversity. It is located deep in the centre of Eurasia, far from the influence of oceanic climate systems. The basin lies at the interface of three great climate regions which extend across east and central Asia: (1) highlands; (2) cold dry mid-latitude steppe; and (3) cold dry mid-latitude desert (Trewartha 1957). The global vegetation classes associated with these climate zones are needleleaf evergreen forest in the alpine zone of the highlands; grasses, other herbaceous forms and shrubs on the steppes, and broadleaf deciduous shrubs in the deserts. Principal soil orders are aridosols in the desert, mollisols in the steppe zone and undifferentiated mountain soils. Elevations range from 4793 m in the Ala Tau mountains to 340 m at Lake Balkash itself. These salient climatic and landscape features shape the dynamics of the Balkash basin ecosystem and regulate its capacity to support livestock and people.

The steep elevation gradient, dropping over 5,000 m in 300+ km between the mountains and Lake Balkash, helps to generate a diverse set of vegetation communities stretching out across the basin. Coniferous forest and alpine meadows form the highest vegetation zones, below permanent snowfields and bare rock. Mountain meadows occur in narrow valleys and on steep to moderate mountain slopes at somewhat lower elevations. Further downslope where the declivities become more gradual, grasslands (mountain steppe) are interspersed with small stands of deciduous trees. Lower steep slopes support dense shrub communities in dry foothills and canyons. Where the mountains meet the plains, typical steppe grasslands dominate. But as rainfall decreases rapidly with distance from the mountains, more xeric shrub elements (*Artemisia*, etc.) intergrade with the steppe grasses, creating a heterogenous mix of communities (semi-steppe *or* semi-desert) which vary with small changes in soil or landscape. Midway between the mountains and Lake Balkash, a vast homogenous expanse of shrub-dominated plains extends to the northern edge of the distribution of grey and grey-brown soils. This shrub-dominated community would be classified as desert-steppe in east Asia; here it is referred to as semi-desert or plain desert. Where the fine-textured grey and grey-brown soils give way to coarse sand and fixed dunes, there is an abrupt shift from small half-shrubs to massive woody shrubs (*Haloxylon*), mixed with bunch grasses, forbs and other shrubs. This zone is referred to as sand desert; it extends to the edge of Lake Balkash. There are numerous wetland sites within the sand desert, a result of a relatively high water table near the bottom of the Balkash basin. The Ili and Topar rivers are the main sources of fresh water for Lake Balkash. The deltas of these streams support a savanna-like mixed community of trees, shrubs and grasses. Near the lake itself, wetland communities, composed of *Tamarix* trees, assorted grasses, reeds and shrubs, are prevalent. (See Chapter 3 for a more detailed description of vegetation.)

Interactions among climate, vegetation structure and livestock populations

Climate data were acquired from six weather stations located within the study area. Thirty years data from the mid-1960s to mid-1990s were analysed to determine long-term climate trends and climatological mean values. These data were also used in conjunction with remote-sensed NDVI data, to assess climate–vegetation interactions.

Seasonal temperature extremes and the value of shrub-dominated ecosystems for livestock in the Balkash basin

Climatic extremes in the Balkash basin are a major constraint for livestock husbandry. The region has one of the higher amplitudes of seasonal temperature variation on earth. The normal range here is 35–40°C or around 60–80°F. The world's highest seasonal temperature variation is found in northern Siberia with a range of over 56°C (100°F). Over the Balkash region, July temperatures average between 20 to 27°C, while January norms are from –7 to –16°C. Temperature data from the study area show that the frost-free period may last nine to ten months in the foothills of the mountains (Uzanagach weather station) and six to seven months in the cooler desert sites, near Lake Balkash (Bereke weather station). However, extremely cold winters with long periods of snow cover are a challenge for livestock. Ice storms or deep snow, which are not unusual in the Balkash basin, can submerge all but the tallest shrubs and cause livestock starvation.

Tall woody shrubs are the mainstay of winter survival for many browsing and mixed-feeding herbivores in the cold-temperate regions of the world. Where tall shrubs are abundant, winter kills are rare. Where shrubs are scarce, heavy snow or ice can decimate ungulate populations. Throughout the northern hemisphere, both wild and domestic herbivores migrate from summer ranges at higher latitudes or elevations to winter ranges where woody shrubs are a major component of the vegetation (Ellis 1970). The northern half of the Balkash basin is covered by tall shrubs on fixed-dune sandy soils. The sloping dunes provide enough micro-topographic relief to allow the snow to melt on south-facing slopes. Winter winds blow snow from some sites, keeping shrub forage open for browsing. This region provides excellent winter range from the point of view of livestock winter survival, and was a crucial component of the migratory strategy of Kazak herders, prior to collectivisation. A reduction in mobility and sedentarisation on more productive, but less protective regions of the basin necessitated the use of fodder (hay) for wintering livestock. The nomadic strategy, utilising the shrub-dominated sand dunes, incurred very little cost for imported fodder, whereas sedentarised livestock in other zones are dependent on fodder throughout much of the winter.

Precipitation patterns

The precipitation gradient

Landscape variations, elevation gradients and the convergence of three major climate regions give the study area a very strong precipitation gradient. This, in turn generates diversity in vegetation communities, the basis of the extensive migratory (or transhumant) system of livestock management developed by Kazak pastoralists. Kazakstan climate maps show that rainfall ranges from 100–50 mm per year on the shores of Lake Balkash to over 1500 mm per year, high in the Ala Tau mountains. Sparse precipitation near Balkash supports desert vegetation except where supplemented by the high groundwater table. Precipitation rises slowly across the plains, reaching 400–500 mm near the base of the mountains. From there, precipitation increases rapidly with elevation, reaching 1500 mm at highest elevations and supporting evergreen coniferous forest in suitable sites (Table 4.1). This regional precipitation gradient is mainly responsible for the changes in vegetation from desert shrub, through shrub-steppe, to steppe, mountain steppe, and sub-alpine forest. Vegetation biomass also increases rapidly with elevation in the foothills and mountains.

Seasonal precipitation patterns: implications for vegetation structure, production and quality

The dramatic seasonal temperature range of the Balkash basin is in contrast to a remarkably low seasonality and interannual variability of precipitation regime, prevalent prior to 1987. Rainfall seasonality has major effects on how ecosystems

Table 4.1 The precipitation gradient: 30-year means and interannual CVs

Climate station & approx. elevation	Arssy 3000 m	Uzanagach 1100 m	Aksenger 700 m	Ay Darla 500 m	Bakanas 400 m	Bereke 400 m
Mean annual precip.	680 mm	453 mm	348 mm	228 mm	190 mm	138 mm
Interannual CV	0.23	0.20	0.22	0.25	0.30	0.25

function and how they can be exploited. In very general terms, vegetation production is maximised by an intense rainy season. The higher the proportion of total annual rainfall that occurs during the growing season, the higher net primary production will be. Some rangeland ecosystems are distinguished by a marked seasonality in precipitation. For example, the great Mongolian steppe to the east of Kazakstan has very strong seasonality with the vast majority of precipitation coming in mid-summer when temperatures are peaking. This pattern provides a high level of production per unit rainfall. The same rainfall pattern occurs in the African Sahel where temperature is not a limiting factor. There, the rainy season is short but intense, allowing for the production of millet and other crops despite the fact that rainfall seems too low for successful agriculture (i.e. 200–300 mm per year). Generally, agricultural crops do best with short sustained rainy periods allowing the crops to reach maturity. Alternatively, livestock may do better when rainfall is more evenly distributed over the entire growing season so that forage nutritional value remains high (Ellis and Galvin 1994).

In the Balkash basin precipitation seasonality is low. Some rainfall/snowfall occurs each month of the year. There is a strong peak of precipitation in April and May at all stations (Figure 4.1) but there are no months with zero precipitation as a long-term average. As pointed out above, this is favourable for forage nutritional content, but it also means that this ecosystem is not particularly productive as a grazing system and even less so where crop production is concerned. Balkash basin forage protein levels remain at or above maintenance requirements (6 per cent for sheep, 5 per cent for cattle) for much of the growing season. In fact, protein levels are quite high (10–15 per cent) in April, May and June, in all vegetation communities except the steppe pastures (Chapter 3). However, no rain-fed cropping is found in the basin in areas with less than about 500 m precipitation: only irrigated

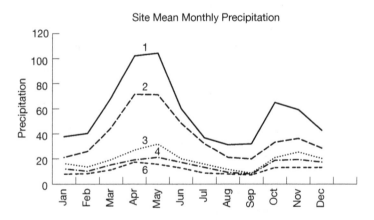

Figure 4.1 Seasonal patterns of precipitation over 30 years, recorded at six weather stations in the study area (see Table 4.1).

agriculture. Contrast this with the situation of the Sahel mentioned above, where crops are grown with only 200–300 mm, all of which comes in thirty to ninety days. In the Balkash basin, only 50 per cent (or less) of total precipitation occurs in the peak rainfall months of April and May.

Plant phenology reflects seasonal temperature and rainfall patterns (Figure 4.1). Steppe, semi-desert and desert communities green-up early (May) as cool season grasses and forbs respond to warming temperatures and early season rainfall. Mountain pastures reach their peak production later in the season (July) as temperatures are slow to warm at the higher elevations. Plants dry out in the desert, semi-desert and steppe pastures during the late summer and early autumn period of low rainfall, but a secondary green-up may sometimes occur later in autumn (October) with increasing rainfall (Figure 4.1). Rainfall seasonality also has implications for vegetation structure. Rain or snow, which occurs during the non-growing season, often infiltrates to relatively deep levels in the soil because it is not used by plants during colder periods. The following growing season, this deep soil water is likely to be exploited by deep-rooted shrubs or trees, rather than by shallow-rooted herbaceous grasses and forbs. This pattern is especially prevalent on coarse-textured sandy soils with high infiltration rates, but it also occurs in regions with finer textured soils. Those ecosystems which receive a significant portion of their annual precipitation during the non-growing season are often dominated by woody shrubs. Where most of the rainfall comes during the growing season, herbaceous plants generally out-compete woody plants for water in upper soil layers. There are many other factors which influence the relative abundance of woody versus herbaceous plants, and annual versus perennial plants, but rainfall seasonality has a major role (Westoby *et al.* 1989). Thus the rainfall regime in the Balkash basin favours shrubs and half-shrubs over grasses and forbs in the dryer portions of the ecosystem.

Interannual variability of precipitation: implications for livestock populations

Year-to-year variations in rainfall determine the interannual stability of ecosystem production. Where interannual variability is great, ecosystem stability is low; where rainfall variability is low, plant production is rather constant or stable from year to year. Variability is measured by the interannual coefficient of rainfall variation. CVs greater than about 0.33 infer high rainfall variability; CVs of 0.20 or below suggest high rainfall stability (Ellis 1994). Interannual rainfall CVs increase with decreasing mean annual rainfall (Nicholls and Wong 1990). Thus many arid and semi-arid grazing systems are subject to high levels of climate-induced instability due to a high frequency of drought years (Caughley *et al.* 1987; Ellis and Swift 1988). These 'nonequilibrial' or 'disequilibrial' ecosystems are prevalent in the dry tropics and subtropics as well as in regions with strong *el Niño* influences (Ropelewski and Halpert 1987; Nicholls and Wong 1990).

The effects of interannual rainfall variability on plant production cascade to herbivore population dynamics. High variability (frequent droughts) causes density-independent mortality in herbivores and oscillating population levels. Single-year 'events' like droughts generally do not result in livestock mortality. However, where an ecosystem is routinely overstocked and degraded, single-year droughts can and do cause mortality and/or poor reproduction (Ellis *et al.* 1987; Ellis and Swift 1988; Coppock 1993). Although single-year droughts do not normally cause animal mortality unless ranges are overstocked, multi-year events do often result in population crashes irregardless of stocking density. This reduces the likelihood that herbivore populations will exceed the nominal carrying capacity of the vegetation long enough to overgraze their plant resources on a continuing basis (Ellis and Swift 1988; Behnke *et al.* 1993). It also reduces the feed-back strength between herbivores and vegetation with the result that animals and plants become more loosely coupled than in more stable systems. Event-driven nonequilibrial ecosystems are typically exploited by a combination of opportunistic strategies including very extensive (nomadic) resource use or 'resource tracking' (Sandford 1983; Westoby *et al.* 1989; Behnke *et al.* 1993).

Low rainfall variability leads to more stable systems where forage availability does not change too much from one year to the next and herbivore populations interact with vegetation in a density-dependent manner. Theoretically, as populations approach the carrying capacity of the vegetation, negative feed-back controls act to reduce herbivore population growth, preventing overgrazing and vegetation degradation. In reality, a variety of factors may intervene to disturb this theoretical balance between forage and herbivore in stable ecosystems. One such factor is supplemental feeding. By supplying fodder, the herder prevents a shortage of forage (or poor forage quality) from exerting its natural effects on livestock populations and productivity. Herd production and growth are then released from the constraints inherent in the natural ecosystem. If populations continue to grow beyond the capacity of range vegetation to support them, then more and more supplemental feed will be required to maintain the livestock population and keep up production levels.

Climate analyses for the Balkash basin revealed a surprising level of stability in rainfall patterns and ecosystem production dynamics from the late 1960s until the late 1980s (see 'Precipitation and vegetation biomass' below). Analysis of long-term climate records from stations near, but outside the study area, demonstrated that this pattern probably extended as far back in time as climate data were recorded, i.e. the 1930s. Even in the driest portion of the basin, interannual CVs did not reach the threshold level of 0.33 for nonequilibrial ecosystem dynamics. In the wetter portions of the study area, CVs were between 0.20 and 0.25, indicating ecosystem stability (Table 4.1). Based on these data, one would classify ecosystem dynamics as equilibrial. However, recent multi-year droughts did occur in the mid-1970s and again in the 1980s. Multi-year droughts frequently cause herbivore mortality in systems where supplemental fodder is not available. Kazakstan livestock population data given in Chapter 5 (Figure 5.1) show no evidence of population reductions in either the mid-1970s or the mid-1980s in conjunction with multi-year drought periods. However both lamb and wool production fell during these droughts. Mortality of adult animals was avoided presumably because supplemental fodder was available to mitigate the effects of drought-induced forage shortages. In fact, livestock data for Kazakstan (Chapter 5) show fifty years from the early 1930s to early 1980s of constant livestock population growth, followed by ten years (early 1980s to 1990) of high but slightly declining populations. The climate data confirm that much of this period was remarkably low in rainfall variability, i.e. equilibrial in its dynamical behaviour. However, Kazakstan livestock populations crashed between 1990 and 1996, in conjunction with severe droughts in 1991 and 1995 during a period of rapidly increasing climate variability.

Regional precipitation trends for the Balkash basin show that rainfall became more variable from 1966 to 1996, with highest peaks and most severe droughts occurring between 1987 and 1996 (Figure 4.2). The two highest rainfall years were 1987, with a sum total of 2629 mm and 1993 with a total of 2703 mm, for all six stations combined. The highest precipitation level prior to the 1980s was 2200 mm in 1973. The driest year was 1991 with only 1142 mm. The next driest year was 1995 with 1314 mm. The driest year prior to the 1990s was 1975 with a sum total of 1574 mm precipitation for all stations combined.

These excessively dry and wet years of the late 1980s and early 1990s caused interannual precipitation variability to increase from low or moderate levels of variability during the 1980s (CV = 0.16–0.25 for five of six stations) to moderate or very high levels (CV = 0.27–0.47 for these same five stations) during the 1990s (Table 4.2). This indicates an abrupt shift from stable ecosystem dynamics to disequilibrial dynamics. Figure 4.2 shows that the shift took place between 1987 and 1996. The steep decline of livestock numbers began in the early 1990s and

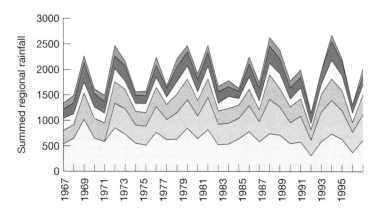

Figure 4.2 Balkash basin long-term precipitation patterns, summed for six weather stations. Data are stacked with highest precipitation station (Arssy) on bottom, lowest station (Bereke) on top.

Table 4.2 Precipitation CVs by time period: 1966–96, 1966–79, 1980s and 1990s

Climate station & approx. elevation	Site 1 Arssy 3000 m	Site 2 Uzunagach 1100 m	Site 3 Aksenger 700 m	Site 4 Ay Darla 500 m	Site 5 Bakanas 400 m	Site 6 Bereke 400 m
Interannual CV:						
1966–96	0.23	0.20	0.22	0.25	0.30	0.25
1966–79	0.23	0.15	0.15	0.18	0.30	0.27
1980s	0.16	0.22	0.21	0.24	0.25	0.23
1990–6	0.27	0.24	0.27	0.36	0.47	0.33

continued through 1998. High climate variability, manifest as severe droughts and intervening very wet years, almost certainly instigated or at least exacerbated the collapse of the Kazakstan livestock population in the 1990s. In the following sections, we suggest that rangeland degradation, along with increasing climatic variability, set the stage for the collapse of the livestock sector. The economic transition then suspended the input flows of fodder and other supplements, leaving the livestock sector dependent on a rapidly deteriorating environmental resource base.

Climate implications for livestock: summary

The Balkash basin has many natural characteristics which make it well-suited for livestock production. The potential effects of cold winters, snow and ice storms are offset by the abundance of shrubs in the ecosystem, especially the tall desert shrub community on the sands south of Lake Balkash. This community provides adequate winter forage for keeping large numbers of livestock, although forage quality almost certainly drops below livestock maintenance levels in winter.

The relatively long frost-free period and the prolonged period of rainfall provide good growing conditions for range forage plants, despite the aridity of a large portion of the ecosystem. Forage nutrient values appear to reflect this, remaining high throughout the growing season. Plant green-up periods differ among plant communities, based on elevation (temperature) and rainfall patterns, with actively growing vegetation occurring somewhere in the region for seven to eight months per year.

The steep rainfall gradient, from mountain to desert, supports a diverse vege-tation community with differing production patterns. Together these communities provide a complete, self-contained transhumant ecosystem, with year-round forage supplies. This should constitute an excellent grazing system, if utilised properly.

The climate data show that interannual rainfall variability was low for most of the last thirty years, but increased dramatically in the 1990s (Table 4.2). This had been a relatively stable ecosystem, one in which livestock husbandry should have been relatively risk-free, compared to many of the world's grazing lands. However, this same stability makes the ecosystem susceptible to over-exploitation when human interventions disrupt negative feed-back processes regulating herbivore numbers. Excessive forage provision can do this and easily lead to overstocking and rangeland degradation. It is very likely that just such a scenario took place during the years of collectivisation, causing degradation of some of the rangeland communities in the Balkash basin. The recent increase in rainfall variability has in effect, shifted the ecosystem from stability toward instability or from equilibrium toward disequilibrium. The severe droughts of 1991 and 1995, when applied within this context of rangeland degradation, could well have pushed a vulnerable system into a downward spiral of lowered livestock production and increased mortality, exacerbated by chaotic economic conditions.

Vegetation patterns and dynamics: evidence of degradation

In the course of this study, field surveys were conducted along a vegetation transect previously established by Dr I. I. Alimaev of the Kazakstan Institute of Pasture and Fodder (Chapter 3). The transect is roughly 200 kms in length, through Dhambul district (*raion*), located in the western portion of Almaty region (*oblast*). It runs from the alpine grasslands of the Ala Tau mountains, through the foothills, on to the grassy steppes. As rainfall declines to the north, the transect follows the gradation of steppe vegetation into the desert-steppe zone, dominated by half-shrubs. The northern half of the transect goes through the sandy fixed-dune desert, dominated by tall *Haloxylon* shrubs. Near Lake Balkash, the transect includes reed, grass and *Tamarix* tree communities prevalent in this zone. The transect includes a series of long-term vegetation sample sites. We selected six of the vegetation sites for analysis of vegetation trends and dynamics using remote-sensed AVHRR–NDVI analysis. These sites represent the most prevalent and important vegetation types, from the perspective of livestock husbandry. During the ground survey of these sites (August 1998), Dr Alimaev identified vegetation type, dominant species and sequences of degradation and recovery at each site, which was located with a global positioning system.

AVHRR–NDVI analysis methods

Remotely-sensed multispectral data have been used for over twenty years to assess and monitor vegetation condition. Numerous studies have shown a strong correlation between spectral reflectance patterns, and plant primary productivity (NPP), biomass, and leaf area index (LAI) (i.e. Tucker and Sellers 1986; Running 1990; Justice *et al.* 1998). Many of these plant–spectral relationships have been documented in arid and semi-arid shrub/grassland regions of the world (e.g. Prince *et al.* 1998).

In this study, Advanced Very High Resolution Radiometry (AVHRR) remote-sensed data, converted to Normalized Difference Vegetation Indices (NDVI) were analysed for the six vegetation sites listed in Table 4.3. The purpose of the NDVI analysis was to identify patterns in seasonal biomass dynamics and biomass time trends over the period 1982–98. We expected that NDVI analysis would duplicate long-term trends in vegetation biomass production and provide evidence on whether or not Balkash basin rangelands were degraded by heavy grazing prior to the livestock population catastrophe.

NDVI data were obtained from three sources. Goddard Space Flight Center, NASA, provided data for years 1981–91. These are bi-weekly composites, registered to the Lambert Azimuth Equal Area map projection. These Global Area Coverage (GAC) data have a spatial resolution of $4.2\,km^2$. Data for years 1992, 1993 and 1995 were obtained from the EROS Data Center, Sioux Falls, South Dakota. These are ten day composites in the form of local area coverage (LAC) with a resolution of $1\,km^2$. No data are available for 1994. LAC data for 1996 through 1998 were obtained from the EU remote-sensing centre, Italy, through Sarah Robinson. These data were treated with the same methods as used by EDC to generate ten-day composites. All data sets were prepared for analysis at the Kansas Applied Remote Sensing Program (KARS), University of Kansas.

The early (1981–91) GAC data sets with a spatial resolution of $4.2\,km^2$ data sets were integrated with the $1\,km^2$ LAC data sets. Both GAC and LAC data are available for 1992. Imagery of the two different types were compared for the same dates during 1992. GAC data were uniformly lower in value, so the GAC data were adjusted to equal the LAC values for the same dates. These adjustments were then applied to all GAC data from preceding years 1981–91. NDVI values were then divided into eleven classes for analysis purposes. We then traced NDVI dynamics at each vegetation site through the period 1982–98. NDVI values were analysed for the first ten-day period (or first bi-weekly period) of the growing season months of May, July and September each year. An average NDVI was calculated for each site, each year. Site NDVI data were compared to annual rainfall figures from the closest weather station (Table 4.1) and to NDVI values for other years to obtain relationships between rainfall and NDVI, as well as NDVI time trends. Estimates of rain-use efficiency were calculated for each site, each year.

Table 4.3 NDVI vegetation site locations, vegetation types and closest weather stations

Weather station	Site 1 Arssy	Site 2 Uzunagach	Site 3 Aksenger	Site 4 Ay Darla	Site 5 Bakanas	Site 6 Bereke
Veg. type	Alpine	Mountain meadow	Typical steppe	Desert steppe	Sandy desert I	Sandy desert II
Location	43° 03.905 / 76° 29.450	43° 06.979 / 75° 50.136	43° 20.621 / 75° 45.120	44° 01.362 / 75° 31.738	44° 29.043 / 75° 26.614	44° 46.166 / 75° 14.043

Precipitation and NDVI dynamics

NDVI values varied from year to year as did the interannual dynamics of precipitation (Figure 4.3). We regressed mean annual NDVI values for vegetation sites, against the annual rainfall figures for the weather station nearest each site. NDVI values for vegetation sites 2–6, reflect the regional rainfall gradient (Figure 4.3). A linear regression model yielded an r^2 of 0.68, suggesting that for the entire study area, the majority of regional variation in vegetation biomass (as represented by NDVI) can be accounted for by differences in precipitation along the rainfall gradient. In this analysis, data from site one (alpine) were not included because climate data from the nearest weather station may not be representative of the site. The closest weather station (Uzanagach) lies at an elevation of about 1100 m whereas the sample site is in the alpine zone, at 2500 m or more. The station at Arssy is about the right elevation, but is over 100 km east of the site.

The data in Figure 4.3 show that over the entire Balkash basin, long-term rainfall patterns are the main determinant of NDVI values. However, on a year-to-year basis, variation in NDVI at any single site is less closely related to local precipitation. Regression relationships between rainfall and NDVI were much weaker for individual sites than for the region as a whole. NDVI dynamics for the dryer sites (4, 5, and 6, with mean annual rainfall ranging from 130 to 230 mm), showed a moderate correlation with rainfall dynamics, while for the wetter sites (1, 2 and 3, with mean rainfall ranging from 350–680 mm). NDVI variation was not closely related to current year precipitation. Patchy rainfall, especially at dryer sites, may cause the vegetation site and the weather station to have slightly different rainfall patterns. This could account for the weak NDVI-rainfall correlations for dry sites.

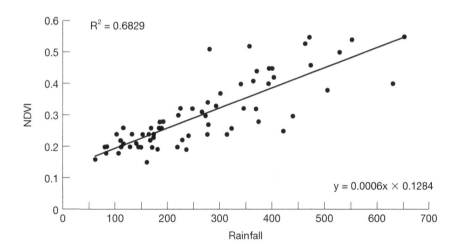

Figure 4.3 Site-level NDVI values for the period 1982–96, plotted against actual precipitation for sites 2–6.

At the wetter, high elevation sites (1 and 2) temperature limits plant production and NDVI dynamics (Yu *et al.* 1999).

At these sites, rainfall may be adequate most years for high levels of primary production; production and NDVI values then depend more on temperature patterns than precipitation. Another complicating factor is the compensatory response of vegetation to high and low rainfall years, discussed below under the heading of rain-use efficiency. This is described in Table 4.4: although the long-term interannual rainfall CVs are moderate to low, the long-term CVs of the NDVIs are even lower, and long-term NDVI variation was not great at any site.

NDVI time trends

NDVI dynamics can be modified by the effects of grazing on vegetation sites. We expected that if grazing pressure was excessive, as livestock population data suggest it might be (Chapter 5), then NDVI values might decline over time as degradation intensified. In fact, there were clear trends of diminishing NDVI values for four of the six sites, between 1982 and 1998. Declines in NDVI values occurred at sites 1, 2, 3 and 5. Site 6 showed a very modest decline, whereas site 4 had a modest increase in NDVI values. This could be interpreted to mean that all sites except site 4 were being progressively degraded over the period of study. If this were the case, we might expect a reversal in the trend after 1993, when livestock numbers began to plummet. This did not occur. Instead, NDVI values continued to decline from 1993 to 1995, with a reversal beginning in 1996 at most sites (Figure 4.4).

The droughts of the 1990s and the increase in climatic variability which took place following the high rainfall year of 1987 could also have influenced the long term NDVI trends. As discussed above, the 1991 and 1995 droughts were the most severe in the rainfall record. The decline in NDVI following these droughts could have been responsible for part or all of the long-term downward NDVI trend. To test these alternative hypotheses (degradation- vs. rainfall-induced declines in NDVI), we examined NDVI trends for the years 1982 through 1993 only. If progressive grazing-induced degradation were occurring, then the negative trends seen in Figure 4.4 should still be observed in the 1982–93 data. If the climate change of the 1990s was mainly responsible for the decline in NDVI, then the downward trend should disappear in the 1982–93 data. The test showed no trends

Table 4.4 Interannual variability of rainfall and NDVIs

Weather station	Arssy	Uzunagach	Aksenger	Ay Darla	Bakanas	Bereke
Vegetation site	1	2	3	4	5	6
Rainfall CV	0.23	0.20	0.22	0.25	0.30	0.25
NDVI CV	0.14	0.12	0.15	0.19	0.13	0.13

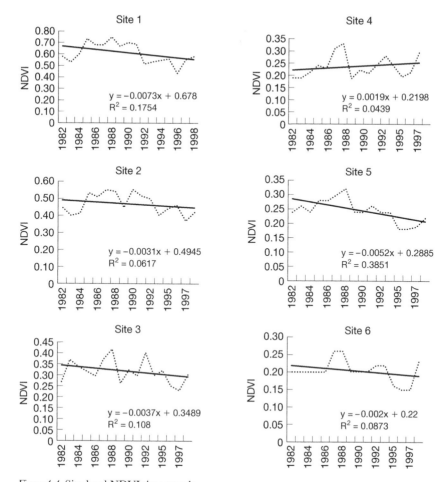

Figure 4.4 Site-level NDVI time trends.

in the 1982–93 data, confirming that climate change, rather than increasing degradation, was the main factor responsible for the declining NDVI trends. However, this does not verify that rangelands were in good condition. It merely demonstrates that no increase in degradation was detectable between 1982 and 1998.

The highest levels of plant biomass production (NDVI) occurred at most sites in the year 1988, following the high rainfall years of 1987 and 1988. The high rainfall years of 1992, 1993 and 1994 following the drought of 1991, did not generate the strong NDVI biomass response seen in 1987–8. Low NDVI biomass values in 1995, 1996 and 1997, followed the drought in 1995. Likewise, relatively low NDVIs were found in 1989, following a decline in rainfall. The 1991 drought, the most severe in the last thirty years, did not appear to have as large an impact as the 1995 drought.

These patterns suggest that the impact of particularly poor or good years carries over for several subsequent years. Thus it is difficult to relate site-level NDVI dynamics to precipitation for that year, in part, because NDVI biomass dynamics reflect multi-year rainfall patterns.

The pervasive effect of the change in climate patterns in the late 1980s can be seen in the regional NDVI patterns (Figure 4.5). The summed NDVI values for the whole region illustrate an increasing trend in vegetation production (NDVI) 1982–8 following a dry period between 1982 and 1984, while climate variability remained low. Afterwards, a downward trend in NDVI continued for a decade, 1988–97, during which climate variability escalated dramatically, accompanied by two severe single-year droughts. This time of reduced productivity and increased variability (1988–97) coincided with the collapse of the livestock sector.

Rain-use efficiency

The quantity of vegetation produced per unit rainfall is a measure of rain-use efficiency (RUE). The more biomass produced per unit rainfall, the higher the RUE. It has been proposed that ecosystem degradation results in a reduction in rain-use efficiency; thus RUEs can be interpreted as an index of ecosystem or range condition (Le Houerou 1984). Declining trends in RUE would presumably indicate that degradation is underway; increasing RUEs would signal recovery.

Recent research explored the use of NDVI to assess rain-use efficiency. Prince *et al.* (1998) used NDVI-based RUEs to search for evidence of desertification in the Sahel. Instead, they found trends of slightly increasing RUEs, which they interpreted as evidence of improving vegetation conditions. They also reported on some general RUE patterns. RUEs tend to increase with decreasing rainfall levels across broad landscape and precipitation gradients; sites with lower average precipitation had higher RUEs. The same pattern occurred at a single site, i.e. RUEs tended to be higher in dry years, lower in wet years.

Rain-use efficiency was calculated for the Balkash basin vegetation sites, each year (RUE = (NDVI/rainfall) × 100). We found, as did Prince *et al.* (1998), that

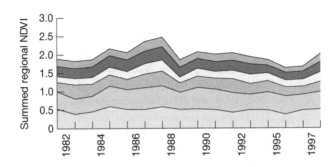

Figure 4.5 Balkash basin NDVI time trend, summed for all sites, 1982–98.

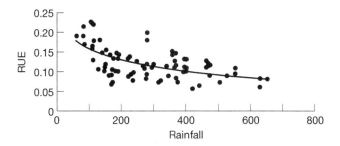

Figure 4.6 Rain-use efficiency all sites, all years, plotted against actual precipitation values.

first, RUE scores declined with increasing rainfall across the region shown in Figure 4.6, and second, for single sites, RUEs tended to be higher during dry years and lower during wet years. The years 1991 and 1995 were very dry and yielded moderate to high RUEs (relative to other years) at all sites. The two wettest years were 1987 and 1993. RUE values were generally low for all sites. This vegetation compensatory response to annual rainfall complicates the interpretation of RUEs as indicators of rangeland condition. On the one hand, low RUE scores could indicate that a site is degraded as Le Houerou suggested (1984). On the other hand, low RUEs occur simply as a response to high rainfall. Thus single-year or short-term assessments of RUE may be confounded by rainfall variability, so it is important to establish an RUE profile for a site, containing data for several years.

We generated three RUE-based tests to evaluate range condition for each vegetation site. First, we calculated the mean RUE for each site. Second, long-term (1982–96) RUE scores were examined for evidence of RUE time trends in the data. As a final test, we measured the sensitivity of site vegetation to rainfall and the stability of vegetation responses to rainfall by plotting site-specific RUE scores against annual rainfall. These tests were applied to each site, with the assumption that low RUE scores, declining trends in RUE and low RUE sensitivity or stability might indicate rangeland degradation.

Mean RUE scores

The RUE scores for sites would be expected to be inversely related to climatological rainfall, based on results of Prince *et al.* (1998) and Balkash regional RUE patterns, presented in Figure 4.6. Thus RUE ranking for sites should be 6>5>4>3>2>1. This was not the case. Site 5 had the highest RUE score and site 3 the lowest. This test indicates that some of the sites might be degraded, relative to the condition of other sites. The range condition ranking of the sites, based on mean RUE scores, is 5>6>1>2>4>3 (Table 4.5).

Table 4.5 Site rain-use efficiency scores

	Site 1	Site 2	Site 3	Site 4	Site 5	Site 6
Veg. type	Alpine	Mountain meadow	Typical steppe	Desert steppe	Sandy desert I	Sandy desert II
Mean RUE score	0.123	0.111	0.101	0.108	0.157	0.127

RUE time trends

RUE scores were plotted over years to determine if increasing or decreasing trends were evident. Results demonstrated that the three wetter sites (sites 1, 2 and 3) had very slight negative RUE trends over the period 1982–96, while the three drier sites (sites 4, 5 and 6) had very slight positive trends, but none of these RUE time trends were significant. When all site data were analysed together, there was a very slight but non-significant negative time trend in regional RUEs (Figure 4.7). If serious rangeland degradation were taking place over the last fourteen years, we might have expected significant declining RUE trends at degrading sites. Conversely, rangeland recovery would have been indicated by strong positive trends. However, the lack of significant trends does not imply that range condition is good; it simply means that the data are so variable that no time trends are detectable.

Vegetation stability and sensitivity to rainfall

Some of the variability in the time trend analysis is undoubtedly due to the compensatory response of vegetation to annual rainfall, obscuring all except strong trends in the data. In order to remove this complication, RUE scores were plotted against rainfall. The resulting linear regressions were interpreted in terms of the sensitivity of vegetation to rainfall and the stability of the vegetation response at

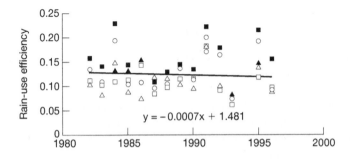

Figure 4.7 Long-term RUE time trend for all sites.

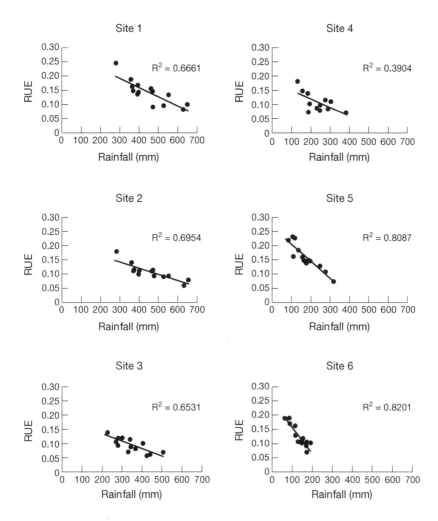

Figure 4.8 Site-level RUE scores plotted against rainfall.

each site. Figure 4.8 illustrates the results of the analysis. It shows that RUE is a function of rainfall at all sites; the higher the rainfall, the lower the RUE.

All site regressions were significant; however, sites differed in the strength and stability of the compensatory response to variation in rainfall. The regressions are interpreted as follows. The slope indicates the *sensitivity* of site vegetation to variation in rainfall. Dry sites were most sensitive, having the steepest slopes. Slopes declined with increasing rainfall but not systematically. Sites 6 and 5 have the steepest slopes, followed by 4, but site 1 had a steeper slope than sites 2, or 3 (Table 4.6). This suggests either that site 1 had an inordinately high RUE slope or that the RUE

Table 4.6 Summary of simple linear regression analyses of rain-use efficiency as a function of rainfall (mm)

Time Period	Site	Intercept	Slope	P	R^2
1982–96	1	0.29	−3.23	0.0004	0.67
	2	0.21	−2.15	0.0002	0.70
	3	0.18	−2.46	0.0005	0.65
	4	0.17	−2.81	<0.0001	0.39
	5	0.26	−5.82	<0.0001	0.81
	6	0.25	−9.07	<0.0001	0.82
1982–9	1	0.24	−2.04	0.0124	0.67
	2	0.17	−1.27	0.0269	0.59
	3	0.15	−1.38	0.0974	0.39
	4	0.13	−1.06	0.5531	0.06
	5	0.26	−5.89	0.0033	0.79
	6	0.23	−7.33	0.0108	0.69
1990–6	1	0.34	−4.39	0.0189	0.78
	2	0.25	−3.07	0.0050	0.89
	3	0.22	−3.55	0.0024	0.92
	4	0.20	−3.91	0.0272	0.74
	5	0.26	−5.93	0.0135	0.82
	6	0.27	−10.99	0.0015	0.94

Note

Regression slope has been multiplied by 10^4. P-values refer to the significance of the regression slope, and describe the probability that the slope of the regression line is actually zero.

slopes of sites 2 and 3 are lower than expected. Site 1 has the least adequate rainfall data set; these data are derived from a station quite far away from the site. Thus it is not unlikely that the slopes for sites 2 and 3 are reasonable, and the rain data give site 1 too steep a slope. If this interpretation is correct, then it can be stated that RUE regression slopes decline with increasing rainfall for intact vegetation communities across rainfall gradients. Slope values range from near −0.0002 at mesic sites to −0.0010 at dry sites.

The coefficient of determination shows the proportion of the variation in the RUE score accounted for by the linear relationship with rainfall. In this analysis the r^2 values are very high for sites 5 and 6, high for sites 1 and 2, slightly lower for 3, and low for site 4 (Table 4.6). This statistical parameter is interpreted to represent the uniformity of the vegetation response to rainfall variability. If the NDVI varies in direct proportion to variations in rainfall, then it can be assumed that the community is in a relatively stable condition and factors other than rainfall have little impact on plant primary productivity and NDVI. If the vegetation community is degrading or if other factors (like excessive grazing) are regularly influencing vegetation responses to rainfall, then the fidelity of the response may differ from year to year, and r^2 values will decrease. In addition, we expect that in communities where water availability is at or near plant requirements, then factors other than rainfall, i.e. temperature, soil fertility, etc., will have more influence on plant primary productivity, NDVI and RUE. So r^2 values should also decline with increasing rainfall across rainfall gradients. The current analysis suggests that the range of values which might be expected for intact, stable vegetation communities is on the order of $r^2 = 0.70$ or greater. This interpretation implies that the vegetation at site 4 may be seriously degraded, and the condition of site 3 diminished, over the period of analysis (1982–96).

There were major changes in the economic system, in climate variability and in livestock numbers in the Balkash basin and the rest of Kazakstan, between 1990 and 1995. For that reason, we re-analysed the RUE data to see if these changes were reflected in RUE trends or patterns. Data for the periods 1982–9 and 1990–6 were analysed separately, with the result that dramatic differences in RUE/rainfall were revealed. The regression relationships were much stronger for the 1990–6 period than for 1982–9. The regressions were not significant at the $P<0.05$ level for sites 3 and 4 during the 1982–9 period, but both were significant in the 1990s (Table 4. 6). The slopes were steeper during the 1990s for all sites except site 5. The r^2 values were all appreciably lower during the 1980s than during the 1990s. The values for r^2 at site 3 rose from 0.39 to 0.92 and the site 4 r^2 value went from only 0.06 during the 1980s to 0.74 in the 1990s. All in all, rain-use efficiency/rainfall increased immensely at sites 3 and 4, and moderately at sites 1, 2, 5 and 6 in the 1990s, as compared to the previous decade.

The increase in rainfall variability during the 1990s (Table 4.2) may account for some of the increase in RUE/rainfall at that time. But the magnitude of the increase in rainfall variability at sites went as follows: 5>1>4>6>3>2. So if the greater variability in rainfall during the 1990s were the prime cause of the increases in RUE/rainfall scores, relative to the 1980s, then we would expect the increase in site scores to replicate the above site sequence (5>1>4>6>3>2).

In fact, site 5 which experienced the greatest increase in climate variability (1980s CV = 0.25, 1990s CV = 0.47) between the decades, actually showed the lowest increase in RUE score during the 1990s. Site 2 experienced the lowest increase in rainfall variability (1980s CV = 0.22, 1990s CV = 0.24), but nevertheless had a moderate increase in RUE scores. Therefore it seems unlikely that the increase in climate variability had much to do with the observed improvement in RUE scores.

The sites with the greatest increase in RUE scores between the 1980s and 1990s are sites 3 and 4, where the increase in climate variability was moderate. These two sites also had the lowest mean RUE scores over the entire period between 1982 and 1996, and the lowest coefficients of determination ($r^2 = 0.65$ and 0.39 for sites 3 and 4 respectively), for the entire period, 1982–96. Our data suggest that rain-use efficiency was impaired at site 3 and especially at site 4, during (and probably before) the 1980s and that RUE scores recovered rapidly and massively in the 1990s. The large increase in RUE scores at these two sites was due, we believe, to a significant improvement in range condition following the crash of the livestock population.

Site 3 and site 4 – typical steppe and shrub steppe communities – were traditionally (before collectivisation) transitional pasture areas, grazed only in the spring and autumn as herders travelled between the winter pastures on sandy desert ranges and summer pastures in the mountains. When pastoralists were settled in livestock collectives, these transitional rangelands were subjected to much longer grazing periods, possibly year-round grazing in some instances. The result has been a serious decline in range condition of the typical steppe (site 3) pastures, and major degradation of the shrub steppe (site 4) pastures.

The steep increase in RUE scores demonstrates recovery, following the release from grazing pressure. The nature of the degradation and recovery of these vegetation communities must ultimately be described in terms of species and productivity changes (see Chapter 3). However the remote-sensed information presented here clearly demonstrates:

- that a major change in ecosystem function has taken place in the last decade;
- the change has resulted in a much improved use of rainfall by typical steppe and shrub steppe vegetation communities; and
- the change coincided with a shift from intensive year-long grazing, to very light or no grazing pressure.

We conclude that this is a clear demonstration of serious rangeland degradation and recovery in some components of the Balkash basin ecosystem. However, this study found little evidence of a degradation–recovery cycle in the other vegetation communities analysed. In general, the sandy desert sites showed little (site 5) or moderate (site 6) increase in rain-use efficiency between the 1980s and 1990s, and other measures of range condition did not suggest that these communities were in poor condition. Site 2, like site 6, showed moderate improvement in RUEs between the decades and could have been in less than excellent condition prior to the 1990s. The rainfall data for site 1 do not allow a rigorous assessment of RUE-based range condition but, based on the analysis carried out, there is little evidence of serious degradation there. These differential degradation patterns may be explained by seasonal grazing patterns. The mountain pastures are only grazed in summer as snow cover makes them unusable in winter. The sandy desert pastures are used only in winter. So serious pasture degradation probably occurred as a result of heavy year-round grazing pressure on the plains vegetation communities. Seasonal grazing was much less detrimental in mountain and desert sites.

Conclusions

The Balkash basin is well suited for extensive livestock grazing. Landscape variations, elevation gradients and the convergence of three major climate regions give the area a strong precipitation gradient. This, in turn, generates diversity in vegetation communities, the basis of the extensive migratory or transhumant system of livestock management that developed here.

The steep rainfall gradient, from mountain to desert, supports different vegetation communities with differing phenology and differing forage plant species. This vegetation heterogeneity provides a complete, self-contained ecosystem, with year-round forage supplies, constituting an excellent year-round grazing system, if utilised properly.

The relatively long frost-free period and the prolonged period of rainfall provide good growing conditions for range forage plants, despite the aridity of a large portion of the ecosystem. Precipitation seasonality is low. Some rainfall/snowfall occurs each month of the year. NDVI analysis showed that steppe, desert steppe and desert communities green-up early (May) as cool season grasses and forbs (ephemerals) respond to warming temperatures and early season rainfall. Mountain pastures reach their peak production later in the season (July) as temperatures are slow to warm at the higher elevations. Plants dry out in the desert, desert-steppe and steppe pastures during the mid-summer period of low rainfall, but a secondary green-up may occur in autumn (October) with increasing rainfall. This temporal phasing means that for 7–8 months, some region in the ecosystem will have actively growing vegetation. The long growing season is favourable for forage nutritional content, with forage nutrient values remaining high throughout the growing season.

The potential effects of cold winters, snow and ice storms are offset by the abundance of shrubs in the ecosystem, especially the tall desert shrub community on the sands south of Lake Balkash. Although forage quality drops below livestock maintenance levels in winter, the desert shrub community provides excellent winter range.

Climate analyses revealed a surprising level of stability in long-term rainfall patterns and in ecosystem production dynamics up until the late 1980s. Even in the driest portion of the basin, interannual CVs did not reach 0.33, indicating nonequilibrial ecosystem dynamics. This ecosystem was, until very recently, one in which livestock husbandry should have been somewhat risk-free, compared to many of the world's grazing lands. However, this same ecosystem stability made the system susceptible to overexploitation. In this case, forage provision allowed long-term overstocking, which in turn led to rangeland degradation in portions of the basin.

Beginning in the late 1980s, the climate regime of the region changed from equilibrial to nonequilibiral in nature, with higher interannual variability due to severe droughts and years with abnormally high precipitation. These droughts came in the early to mid-1990s while economic support for the livestock industry was waning. The droughts probably reduced forage production during summer, causing livestock to enter the winter with less accumulated fat and in poor condition. Range condition and livestock condition were probably worst on those collectives located on the steppe and shrub-steppe vegetation types where range degradation was greatest (Table 4.6). If imported winter fodder were plentiful, this would not have been an overwhelming problem. However, fodder had become scarce and expensive, causing livestock body condition to have deteriorated rapidly. Combined with the economic shocks of this period, this led to large-scale destocking.

The non-recovery of pasture production in the good rainfall years of 1992–4 (Figure 4.5) suggests that long-term overgrazing had taken a severe toll on the resilience of pastures, further compromising the situation of the Balkash basin pastoralists. Thus the evidence presented here indicates that rangeland degradation, along with increasing climatic variability, was an additional factor in the collapse of the livestock sector. Had the increased climate variability occurred at another time when economic conditions were more robust, the results might have been prevented by increasing levels of inputs. Instead, the economic transition suspended the input flows of fodder and other supplements, leaving the stressed livestock dependent on a degraded environment and the pastoralists without the state-sponsored support they had come to depend upon. The combined result of increased climate variability, range degradation and economic failure is shown in the graph of plummeting livestock numbers in Chapter 5.

Recovery of the Balkash basin livestock industry will depend on a variety of factors. The evidence presented here suggests that climate and range condition, along with range use patterns and economic circumstances, will influence the rate and magnitude of recovery and the sustainability of emerging land-use systems.

Acknowledgements

The authors would like to thank Pete Weisberg for assistance with statistical procedures; Randy Boone for helping compile and analyse remote-sensed data; Sarah Robinson for providing us with AVHRR–NDVI data and Ilya Alimaev for introducing us to the Balkash basin and for sharing his lifetime of experience and knowledge of the ecology and socio-economics of the region.

5

RECONFIGURING PROPERTY RIGHTS AND LAND USE

Roy H. Behnke Jr.

Introduction

For Kazak livestock producers, the demise of communism has proved to be nearly as traumatic as its creation. Kazakstan supported over thirty-five million sheep in the late 1980s. By 1999 – following independence, economic reform and privatisation – less than ten million sheep remained. Nearly three-quarters of the national flock has disappeared in the last decade. In the twentieth century, only forced collectivisation under Stalin has had such a devastating impact on stock numbers and rural prosperity.

Figure 5.1 illustrates the extent of the crisis. Prior to the current period of problems, which began in earnest in 1994, sheep and goat populations in Kazakstan suffered two setbacks in the last century – in 1916–20 as a result of rebellion, revolution and civil war, and again in 1929–33 as a result of collectivisation. The causes of the earlier crises are reasonably well understood. The destructive consequences of violent chaos are readily appreciated, and collectivisation was brutal because it was meant to be so. During collectivisation the communist authorities intentionally set out to eliminate a nomadic way of life and a pastoral elite that were deemed to be primitive, inherently anti-Soviet and a threat to the state. The loss of human life and livestock wealth in the collectivisation process was, consequently, higher in Kazakstan as a proportion of the population than anywhere else in the Soviet Union (Olcott 1995: 185).

While admitting that 'market corrections' were inevitable and that certain segments of the population would suffer, few advocates of reform in the early 1990s predicted that it would destroy productive assets and impoverish people on a Stalinist scale. Something has gone wrong with the reform process in the pastoral areas of Kazakstan, and the immediate purpose of this analysis is to identify what went wrong.

Based largely on secondary sources and administrative records, the opening sections of this chapter discuss the privatisation of livestock and the laws that govern the privatisation of land in independent Kazakstan. The middle sections of the

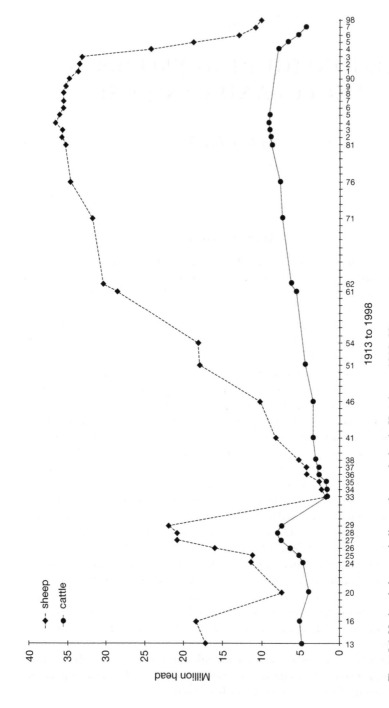

Figure 5.1 National changes in livestock population in Kazakstan, 1913–98.

chapter shift from an examination of national trends and policies to an analysis of specific instances of reform at the local level. These sections present the results of field research on the privatisation of four former state or collective farms along the study transect. Just how reform took place on these farms was, as we shall see, an idiosyncratic process indirectly linked to national legislation and policy. The closing sections of the chapter examine the new forms of commercial pastoralism and pastoral land use that have begun to emerge following reform. Successful post-reform livestock businesses are organised around extended families and frequently draw upon non-pastoral financial resources. When flocks are small, these operations tend to be sedentary and village-based, but larger flocks are generally managed on a migratory basis. The closing section of this chapter discusses the kind of tenure arrangements that are needed to support the mobile, family-based husbandry systems that are spontaneously evolving in response to commercial incentives.

Methodology

The nature and extent of reform in the countryside was investigated in a study carried out in 1998 and 1999 on a sample of former state and collective farms. Field research was conducted for a total of three months in the summer of 1998 and the winter of 1999. Work in 1998 included several weeks of interviewing pastoralists and administrators in Dhambul *raion* combined with a brief reconnaissance of the entire study transect from the western tip of Lake Balkash to the Ala Tau Mountains. Research in the winter of 1999 concentrated on collecting data on the reform of three former Soviet farms situated along the study transect in Dhambul *raion* (district). In the course of this work it became clear that land tenure conditions varied according to proximity to urban areas as well as along a north–south gradient. A fourth peri-urban study site was therefore visited, though it was not possible to examine it as intensively as the other case study farms. Interviews were conducted with shepherds, private farmers, and provincial, district and local officials. Special efforts were made to contact large flock owners who were potential or actual private land-owners.

The immiserising transition

Figures 5.2 and 5.3 clarify the process of livestock privatisation in Almaty *oblast* and Dhambul *raion*, respectively the province and district that correspond most closely to the study area. At both administrative levels, changes in animal numbers broadly parallel those for the national herd and flock (Figure 5.1): that is, after several years of small but steady losses in the late 1980s, populations drop sharply in the early 1990s. Figure 5.2 shows, however, that sheep, cattle and horse numbers did not decline at similar rates. According to Figure 5.2, the numbers of all kinds of domestic stock fall consistently between 1990 and 1996, to rebound suddenly in 1997 when the statistical services realised that privately-owned livestock were

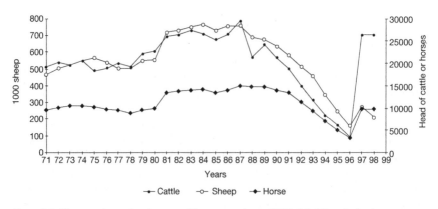

Figure 5.2 Changes in cattle, sheep and horse numbers, 1971–98, Dhambul *raion*.

Note

The jump in livestock numbers between 1996 and 1997 reflects changed methods of data collection rather than altered population levels; see text for details.

becoming too numerous to ignore and included these animals in their estimates. Once these corrections are made, it is clear that in Dhambul *raion* large types of stock have suffered fewer losses than small types of stock in the 1990s. In 1998, privately and collectively owned cattle and horses either slightly outnumbered (for cattle) or were slightly fewer (for horses) than state-owned herds prior to independence. Sheep and goats, on the other hand, had not fared so well, the total flock in 1998 being less than a third of the state-owned flock just before independence. Even in comparison to other forms of pastoral wealth, the sheep flock has suffered disproportionately high losses in the transition to a market economy. The decline of the sheep industry does not merely reflect wider problems in the rural livestock economy; it leads the way.

Evidence on how these losses occurred is provided in Figure 5.3, an examination of the changing balances of private and collectively owned sheep in Almaty *oblast*. By the close of the 1990s, sheep ownership in the province had shifted decisively from the state and large collective sector into private hands. Private owners held 10 to 15 per cent of the provincial flock in the 1970s and 1980s, 75 per cent by 1997, and almost certainly hold a higher proportion of the total today. The privatisation of livestock holdings is therefore an accomplished fact.

But it could hardly be deemed a success. Especially for sheep, privatisation has been achieved less by reallocation than by the simple liquidation of collective flocks. Until 1994 when stock numbers began to decline sharply, the absolute number, as well as the proportion, of privately owned sheep increased. After 1994, however, the number of privately owned sheep has not grown and the increasing proportion of private sheep in the provincial flock is largely attributable to the disappearance of collectively owned animals. The result is an apparent paradox: private flocks expanded in the late 1980s and early 1990s when the legal

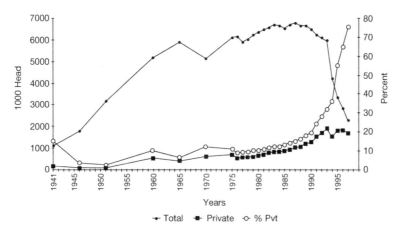

Figure 5.3 Almaty *oblast* sheep numbers, 1941–97: private holdings as percentage of total.

foundations for privatisation were weakest, and expansion stalled in the mid-1990s as the reform process gained momentum and improved legal recognition.

A combination of factors explains this arrested pattern of commercial development. Private pastoral production, like private agriculture in general, occupied a restricted but privileged niche within the late Soviet economy. Based on lucrative internal Soviet markets and sustained by both cash wages and in-kind (often covert) subsidies from the state system, private flocks prospered. But they did so as dependants of the larger socialist system, not its antithesis, and when that larger system started to crumble, they too began to waste away. A small number of private producers could forgo or find substitutes for the supports that the state had once provided, and they have prospered in an increasingly liberalised commercial environment. But most could not. As a consequence, recovery has been retarded and is restricted to a small circle of entrepreneurial producers with rather special circumstances.

By 1995 new legislation provided the legal basis for such a realignment of property rights and ownership patterns. However, it has been in the interests of the ex-Soviet rural elite to either preserve the old system or to wind it down slowly, and the central government has been either unwilling or unable to challenge this position. As a consequence, economically dysfunctional, nominally privatised public sector agricultural enterprises have lingered on, eating into their capital without hope of renewal and selling off parts of their operation on a piecemeal basis. Asset redistribution is occurring – reluctantly.

As the preceding account suggests, the reform of Kazakstan's sheep industry has passed through two relatively distinct phases. In the first phase, from independence in 1991 until 1995, the Soviet system of economic exchange dissolved in chaos; in response, Kazakstan created its own currency and removed most price controls and subsidies on agricultural commodities. By the close of this period Kazakstan had a thoroughly market-oriented if debilitated agricultural economy. These developments form the background to the present study and are briefly reviewed in the next section of this chapter.

By 1995 Kazakstan also had – at least on paper – the rudiments of a new system of private property rights. The enactment of these laws opened up the second phase of the reform process. Having transformed the system of agricultural exchange, the potential now existed to reorganise agricultural production by redistributing the ownership of productive assets. But widespread property redistribution would necessarily cause major institutional changes in the countryside. The main body of this analysis looks at the extent of these changes, how they took place, or why they did not occur in one rural *raion*.

1991–5: market loss, hyperinflation and the sheep industry

The Central Asian republics, Kazakstan included, were particularly vulnerable to the disintegration of the Soviet economy. In addition to the routine Soviet problems of inefficiency, resource misallocation and primitive technologies, the republics operated specialised, export-dependent economies that left them ill-prepared either to stand alone or to retreat into economic self-sufficiency (Pomfret 1995). As late as 1995, Kazakstan's prime minister estimated that 80 per cent of the country's economy remained tied to the Russian Federation (OECD 1995: 129). Unfortunately, many of these ties 'made sense only in the wider context of the Soviet economy (if then), as enterprises produced for captive markets or participated in a Soviet-wide processing chain' (Pomfret 1995: 32). When Soviet markets and marketing chains collapsed in the early 1990s, many Central Asians found themselves making products that nobody wanted.

This was the fate of Kazakstan's wool producers. In common with much of Soviet industry, Kazakstan's wool industry was organised to produce high volumes of output that met minimum quality thresholds (ADB 1997: 23). Much of this wool was used in Russia for the manufacture of items such as blankets, army uniforms and better-quality carpets. When Kazakstan was forced onto international markets in the early 1990s, there was an oversupply of wool of this quality, demand was weak and Kazak producers could expect to sell less wool at a lower price than they had been accustomed to in the Soviet period. In the event, even these modest expectations were not met. By 1995 the Chinese had replaced the Russians as the principal importers of Kazak wool, but export volumes had fallen to about a tenth of what they had been prior to independence (ADB 1997: Table 8 and Chart 26). Moreover, the centralised, Soviet-era wool sorting and cleaning infrastructure had collapsed. Much of the wool that was now leaving Kazakstan was being sold ungraded and unwashed at knock-down prices to small traders immediately after the close of the shearing season when prices were lowest (ADB 1997). Wool had once been a valuable product; now it was sheared largely to get it off the sheep's back, the value of the wool being little different from the cost of shearing (see Chapter 8).

Kazak sheep owners had previously harvested a combination of meat and wool. Now they could only profit from sheep by slaughtering them. Although the evidence is circumstantial, the changing composition of the national cattle herd at this time suggests that these changes in the value of sheep output encouraged higher sheep slaughter rates and depressed flock sizes. Male cattle, like sheep kept only for meat, offer only terminal products derivable from their carcass. Cows, like mixed wool and meat-producing sheep, provide both live-animal offtake (calves and dairy produce) and, eventually, a carcass. Between 1985 and 1995 Kazakstan's total cattle numbers declined by 25 per cent while the stock of milk cows declined by only 2 per cent. In an economic crisis, cattle owners sacrificed animals that had only a slaughter value and retained dual-purpose stock. If shepherds applied similar criteria to their sheep flocks, then the loss of wool markets almost certainly contributed to declining sheep numbers, to an extent that is difficult to determine since so much else was changing at the same time.

Hyperinflation and currency devaluation also played a role in decimating the national flock. Inflation peaked at a monthly rate of 55 per cent in November 1993; savings in local currencies that were worth one US dollar at the beginning of 1993 were worth about two cents at the end of 1994 (ADB 1997: Table 1). As money lost its value, Kazakstan experienced a severe liquidity crisis and financial gridlock; no one could pay their debts because none of their creditors could pay them. Thus, in 1994 farms owed their workers about 1.5 billion tenge in back wages; but the dairy and meat processors alone owed farm producers 545 million tenge, and were themselves owed 422 million tenge by the wholesale and retail outlets (OECD 1995: 133, 135). As cash became increasingly scarce, trade was conducted in barter, which offered the additional advantage of avoiding bank transfers and the seizure of money by creditors (Carana 1996: 11). Many farms simply sold off their productive assets in order to pay their debts.

These developments were especially significant for the sheep industry because sheep – plentiful, portable and of appropriate unit value – became a substitute currency, used for paying everything from wages to electricity bills. Unfortunately, sheep were not a stable currency. Most of those who were paid in sheep did not want to breed them or eat them; they wanted cash, and quickly dumped their animals on the market in order to obtain it. As more sheep were offered for sale, their value declined, forcing sheep owners to dispose of ever more animals in order to meet their obligations or recoup the purchasing power of their lost salaries. Over a quarter of the national flock disappeared in 1994 alone.

By 1995 the rate of both inflation and of currency devaluation had fallen off. Kazakstan had also abandoned, or been forced to abandon, a command economy with centrally planned purchase orders, production targets and prices. What the economy now needed, and did not yet have, were new forms of agricultural production that would stabilise the situation by taking advantage of the new commercial environment.

Privatisation after 1995: legal framework and general trends

The dissolution of a communist and the creation of a capitalist system of production required the reallocation of three different kinds of pastoral assets: land, livestock and the infrastructure and equipment associated with livestock production. Between 1990 and 1995 the legal basis for such a reorganisation was uncertain in Kazakstan. With about twenty separate pieces of significant legislation, and in the absence of adequate definitions of critical legal terms, 'the farm privatisation legislative framework evolved piecemeal rather than as a crafted design with a foreseen outcome' (ADB 1996a: 25). By 1995, however, changes in the constitution, fresh presidential edicts and the issuance of the Civil Code had significantly clarified the legal situation, which remained basically unchanged until early 1999 when field work for this study was completed. In brief, the legal framework for farm privatisation was as follows:

- In 1995 private land ownership was explicitly recognised, though it was restricted to house, garden and vegetable allotment (*dacha*) plots that were not technically classified as agricultural land. For agricultural land, including land owned by peasant/family farms, the most secure form of tenure was provided by 'permanent use' rights – a concept that had evolved to include the right to sell or transfer property rights and by 1995 was virtually indistinguishable from outright ownership (ADB 1996b: 15, 25).
- Legislation to create peasant farms – i.e. unincorporated family farms – was introduced before independence in 1990. Under this legislation private farms could be created directly from property held by state farms. By 1999, however, the vast majority of private farms were being formed through a two-stage process that first involved the dissolution of the state farm, as described below.
- Introduced in 1992, legislation to privatise state farms rested on the issuance of land share certificates. In theory, the value of these paper shareholdings was determined by equitably dividing the land of a state-owned agricultural enterprise among the people entitled to it – workers, managers and service personnel. The new shareholders then met in a general meeting and, typically, reincorporated their organisation as a private entity such as a joint stock company or production cooperative.
- Individuals wishing to leave these newly privatised cooperatives were free to do so. To create a peasant farm or a smaller farming enterprise, they had to redeem their land share certificates for registered titles to specific, demarcated plots of land. They were also entitled to claim their share of the collectivity's non-landed assets, including livestock and agricultural equipment.

In sum, the law encouraged the privatisation of agricultural land and assets in two phases. In the first phase, state (*sovkhoz*) or collective (*kolkhoz*) farms are transformed into legally independent commercial entities. In a potential second wave of reform, these newly privatised businesses were subjected to further rearrangement and possible dismemberment, as individuals or small groups opted to leave the larger organisation.

The speed and depth of these two reform processes tended to be inversely correlated. In comparison to the auctioning of urban industrial organisations, the privatisation of state farms went very quickly because ownership was 'simply . . . transferred to the employees and consequently no major ownership change-over [was] brought about' (Haghayeghi 1997: 327). One observer declared the farm privatisation process 'almost complete' with the re-registration in 1996 of 93 per cent of all state farms as private entities (ADB 1996b: 1). However, for many state farm employees, the acquisition of private land rights went no further than receiving and then relinquishing their land certificates, usually to their former state farm directors, who now served as the directors of restructured private enterprises. While the new private cooperatives operated in a transformed macro-economic environment, from the perspective of the individual employee, much remained the same. Senior management had likely survived the reshuffle, the cooperative probably had a new name but had acquired no new assets, lost very few members, and retained most of the financial liabilities it had accumulated before privatisation. In these cases, 'Agricultural production assets have not . . . changed their economic and social status' (ADB 1996a: 29). Nor, in all probability, had the new operation become notably more efficient: in 1995 the Ministry of Agriculture estimated that 80 per cent of farming enterprises were still unprofitable (ADB 1996a: 9).

It remains to be seen whether a second wave of privatisation will now lead to the formation of many private, family farms. This has not been the case thus far. At the national level, by the close of 1995 on average across Kazakstan less than one peasant farm had been created from each *sovkhoz*, which had spawned on average 2.6 'non-state agricultural enterprises' (ADB 1996a: 29). Using a different measure of change at a later date, privatisation seems to have progressed somewhat further in Almaty *oblast*. By the close of 1997, 22 per cent of all arable land in the province was under the ownership of 14,066 private farmers or small cooperative operations, up from 1 per cent of sown land before independence (Sabeko 1998). Moving down to the district level, by early 1999 Dhambul *raion* contained 650 private farmers, whereas none had existed in 1990. No information was obtained on the proportion of district land held by these 650 farmers. However, by 1999 within the four communities examined at first-hand in this study, private farmers and small-scale collective enterprises held less than 7 per cent of the land of their former state and collective farms.

The evidence presented thus far points to the thorough 'destatisation' of agricultural land followed by a very modest level of internal farm reorganisation. If this is the case, rural Kazakstan may be a long time building commercially viable production enterprises that can fully exploit the new marketing system.

Economic pressure and institutional inertia in the countryside

Figure 5.4 summarises the main characteristics of the Dhambul *raion* land tenure system in the late 1980s. By the end of the Soviet period, farms were huge, state farms averaging 82,000 ha for Kazakstan as a whole in 1994 (Green and Vokes 1997). Virtually all of the land in the old administrative district of Dhambul *raion* was controlled by fifteen farms, while nineteen farms controlled the bulk of agricultural land within the enlarged district that resulted from post-independence boundary changes. Fifteen of the farms in the 'new' Dhambul *raion* were state farms (*sovkhoz*) on which assets were owned by the state and workers were salaried employees. Two of the farms were collectives (*kolkhoz*) nominally owned by the workers themselves, and the last two farms were an agricultural research institute and its experimental farm.

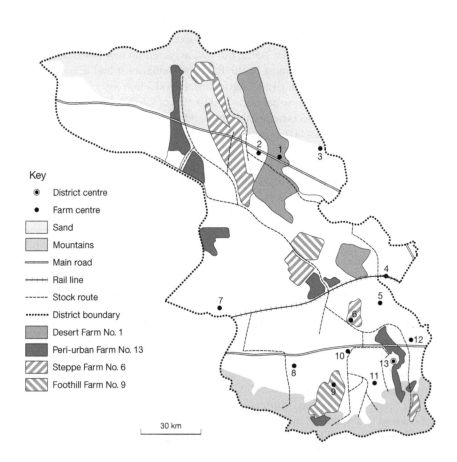

Figure 5.4 Dhambul *raion*: land ownership by case study farms.

Irrespective of these differences in legal status, farms were of much the same size and territorial organisation. The district contained one specialised camel farm and one horse farm. With these exceptions, all large farms in the district were involved in mixed arable farming and sheep husbandry and possessed sheep flocks ranging in size from about 25,000 to 60,000 head (see Table 5.3 on page 101). Farms in the district were on average made up of about five territorially-distinct sections. As is evident from Figure 5.4, these sections tended to be arrayed along a north–south axis, from high to low elevation and across the grain of ecological variation. About half of the farms owned parcels scattered from the sands in the far north of the district to the mountain pastures in the far south. This arrangement gave each farm seasonal access to a variety of pasture types linked together by an elaborate network of livestock trek routes (Figure 5.4). By allocating a variety of land types to each farm, the scattering of parcels also facilitated the combination within farms of arable farming, fodder production and animal husbandry.

Despite the dispersed pattern of land ownership, each Soviet farm had its main population and administrative centre in a particular ecological zone. As the Soviet system fell apart, these central villages/administrative centres lost the ability to exploit peripheral parcels of land that they had once controlled. Villages/farms thereby acquired distinctive geographical centres of gravity – core areas made up of land around central villages and nearby sections of critical economic importance. Where villages are located has therefore determined which natural resources the villages can still exploit, and this, in turn, has influenced their prosperity and their response to tenure reform. The four farms examined in this study are situated in widely separated ecological zones, at varying distances from Almaty city. Their different histories illustrate some of the diversity that has arisen as rural communities with different resource endowments respond to national policies and land tenure legislation.

Table 5.1 and Figure 5.5 summarise some of the results of this work.

Figure 5.5 contains no surprises; it documents the deterioration of the number of sheep held by the four large, privatised cooperatives, heirs to their respective state or collective farms. At least with respect to their inability to hold onto sheep, these cooperatives were fairly typical. Table 5.1 is more revealing; it compares the rate and methods of property reallocation engaged in by the different cooperatives. According to this table, even for large, high-profile enterprises governed by a single district administration, the approach to privatisation varied markedly from case to case.

According to Table 5.1, some farms began land redistribution in 1992; others were still thinking about it 1999. Similarly, some cooperatives retained all their barns and stock shelters, while others sold them to whomever was interested, or allocated them to cooperative members – sometimes with and sometimes without accompanying rights to the land that the buildings sat upon. Likewise, two cooperatives in this sample did not redistribute sheep to their members; two other cooperatives did, but each did so in its own way. One held a huge, one-off distribution of all of its sheep and associated infrastructure in 1997; the other

Table 5.1 The privatisation of four state or collective farms in Dhambul raion

Farm pseudonym and date of destatisation	Sovkhoz or kolkhoz area in ha	Private land – ha end of 1998	Private land – % of total	Number of private farms and their registration dates	Methods of livestock redistribution	Methods of allocating barns	Total 1998–9 flock size – % of pre-reform state flock*
Steppe Farm 1996	87,000 sovkhoz	50	0.1	1 Registered in 1995	No redistribution	No individual ownership	24
Foothill Farm (rural) c 1996	43,000 kolkhoz	98	0.2	7 All registered in 1998	Sheep allocated to members in 1997	Barns but no land rights; allocated in 1997	31
Peri-urban Farm 1996	102,000 kolkhoz	5,597	5.5	85 88% registered in 1997–8	No redistribution	Cash sale with land rights	47
Desert Farm 1995–6	103,000 sovkhoz	15,588	15.1	16 Registered from 1992 to 1998	Allocated upon exit from coop	Barns with land rights; allocated upon exit from coop	20

Note

* Estimates of current flock sizes are based on district-level official statistics that are unreliable but probably provide an accurate impression of the relative performance of different farms. Both field observations and the municipal authorities at Desert Farm support an estimate of total flock size that is about half of the official figure for that farm. The estimated rate of flock survival on what used to be Peri-urban Farm – 47 per cent – is almost certainly too high, it being difficult to separate animals owned by farm members or ex-members from those owned by town residents in general.

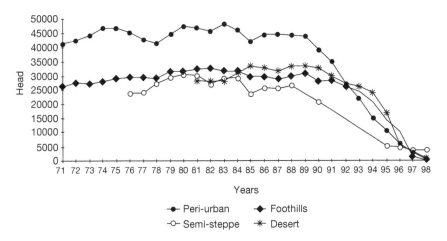

Figure 5.5 Sheep populations, 1971–98: collective, state and cooperative animals.

distributed sheep gradually as part of the severance packages that individuals received whenever they left the cooperative.

The pathways to privatisation taken or not taken by these communities do not add up to a consistent picture. For example, the rural foothill cooperative that successfully and equitably redistributed its sheep assets, strenuously resisted attempts by its members to privatise the arable land needed to engage in grain production. Conversely, the peri-urban foothill cooperative that encouraged its members to obtain private land rights was not prepared to allocate any other productive assets to them. And the steppe cooperative did none of the above, although the managers admitted that they were under increasing pressure from the district authorities to do something.

There is, moreover, no simple correlation between the location and the commercial success of the larger cooperatives in the sample. Financially the two least successful large cooperatives – the desert and the peri-urban case studies – differed markedly with respect to their geographical location, ease of market access, and rates of flock survival. Table 5.1 suggests neither a uniform pattern of change nor an obvious formula for successful privatisation. What the table does suggest is that cooperative managers have been left with considerable discretion to interpret the privatisation laws as they see fit within their own organisations. That they still have this power reflects their success in the national debate on the nature of privatisation in the countryside.

Cynthia Werner identified a range of attitudes in the early 1990s among the Kazak elite toward the privatisation of agriculture (Werner 1994). At one extreme were the 'free marketeers' urging radical reform and minimal state interference along lines advocated by neo-classical Western economists. At the other end of the spectrum lay the 'neo-collectivists' who advocated the continued strong involvement of the state in agricultural production. The backgrounds of these two groups were very different. The free marketeers tended to be urban intellectuals with no direct involvement in agriculture; the neo-collectivists were 'leaders of agricultural workers' unions and the local managers of agricultural production (at the *kolkhoz*, *sovkhoz*, *raion* and *oblast* level)' (Werner 1994: 298). Already in the early 1990s, in Werner's estimation, 'as the privatisation process unfolds, it is becoming clear that the dominant state position is gradually shifting from the "free marketeer" . . . to the "neo-collectivist" perspective' (Werner 1994: 301). Besides, she observed, 'it is the "neo-collectivists" not the "free marketeers", who controlled a powerful resource base', that is, agricultural production (Werner 1994: 301). She might also have added that the neo-collectivists – former members of the Soviet rural elite – had the most to lose in any radical break with the old order.

But they did not lose. In the complex equation that explains the varied pattern of privatisation on individual farms, the 'neo-collectivist' perspective is the ideological constant. At least on their own ground in the countryside, the neo-collectivists reign supreme, consistently favouring collective solutions and instinctively opposing institutional change. In this ideologically constrained environment, change is driven by economic variables that disturb institutional inertia. Either financial duress or attractive economic opportunities can provide the impetus for change, but the changes tend to be the minimum that is required and to preserve the institutional status quo whenever possible. When the economic signals are either weak or mixed, nothing happens.

The following sections examine the fate of the four collective/state farms in Table 5.1, each subjected to different economic conditions. We begin with an examination of a community, Steppe Farm, in which economic incentives and disincentives have been more or less equally balanced and little change has occurred in the formal tenure system. Steppe Farm gives us a glimpse into rural Kazakstan's Soviet past, a clue as to how farms functioned before they evolved along divergent pathways in response to varied economic stimuli. We then turn to a story of relative success, Foothill Farm, for an example of cautious change driven by positive commercial incentives. Finally, we consider two farms (Desert and Peri-urban) that have suffered economic collapse since reform and where institutional change has occurred under duress.

Stasis and paternalism – Steppe Farm

Steppe Farm emerged as a *sovkhoz* in its own right in 1976, was re-amalgamated into an adjacent farm in 1988, re-emerged in 1990 as a state cooperative, and became a private cooperative in 1996. At least initially, privatisation could reasonably be viewed as simply the latest instalment in a process of continuous but superficial farm reorganisation. Privatisation transformed the economic conditions under which the farm operated – state support ceased and the cooperative became responsible for its own production and marketing arrangements. But reform did not necessarily affect the farm's internal organisation, since it was already an independent cooperative, and the local response was conservative. By 1999 there had been neither livestock redistribution nor land reallocation.

Private livestock husbandry existed under the protection and at the sufferance of the cooperative. Both pensioners and the unemployed owned private livestock but prosperous flock owners tended to be at least nominal cooperative employees. On a pattern that seems to have persisted from the Soviet period, the accumulation of a substantial flock depended on holding a job that provided a convenient platform for private accumulation. In 1999 there were about two dozen non-managerial positions at Steppe Farm ideally situated to foster flock growth – about a dozen positions guarding rural buildings and a similar number of herding and shepherding positions. These jobs provided a place outside the village – and hence outside the ring of heavily grazed pastures around the village – where an employee could live and a flock could prosper. Guarding jobs routinely involved someone living in a cooperative dwelling that they would, in any case, have wanted to occupy. The duties of a guard were slight and did not interfere with other income-generating activities. Shepherding for the cooperative was more onerous but offered a better salary and collateral benefits. The private livestock holdings of cooperative shepherds averaged around fifty head of sheep plus a few cattle and horses. With few exceptions cooperative shepherds kept these animals with them, so that their private stock migrated seasonally at cooperative expense and were shepherded at little extra cost to their owners. Cooperative shepherds also had informal but implicit access to the fodder and feed supplements intended for the cooperative's animals.

Some assistance from the cooperative was available to village residents irrespective of their employment. Occasionally tractor drivers diverted cooperative fodder into private stores. The cooperative explicitly allowed village residents to cultivate potato gardens for the family and harvest natural pastures, but not sown hay fields, for their private flocks. As on all the farms studied here, village houses and house plots were allocated to their occupants. Whether it was called a collective or a cooperative, the farm was there, everyone agreed, to help its members. *De facto* private use coexisted with *de jure* collective ownership of many critical resources.

The farm controlled a mix of poor- and good-quality pasture and arable land. The intermediate nature of its resource base meant that Steppe Farm experienced neither spectacular farming success nor failure of the kind that had forced the pace of change on other farms. The farm was prosperous enough to have no large outstanding debts, for instance, but poor enough to continue to pay its taxes in grain. If it had the largest collective flock of all the farms sampled in this study, it also had the largest workforce. The demands on the flock were such that the cooperative – like the poorest of private flock owners – could not fatten animals for the lucrative winter and early spring market. By late winter the cooperative had no animals left to fatten because they have been forced to sell all of the previous spring's male lambs in order to cover operating costs. Despite conscientious management, Steppe Farm limped along maintaining a level of welfare spending and a generous employment policy that undermined its profitability.

ROY H. BEHNKE JR.

From collective to producer cooperative – Foothill Farm

Foothill Farm was founded at the turn of the century by Russian peasants on land used seasonally by Kazak pastoralists. Given its relatively long history and favourable resources, the farm had a wide range of economic options: grain and potato production (the legacy of the original Russian settlers), sheep husbandry (the economic background of its Kazak members), large-scale mechanised grain production on the steppes and semi-desert (thanks to the Virgin Lands Programme), and a dairy operation. For the farm management, reform has been a process of sorting out which of these enterprises were commercially viable, retaining those that were and disposing of those that were not.

After privatisation farm managers decided that extensive grain production in low rainfall areas would not make money at market prices, and abandoned this enterprise and any further attempt to maintain a presence on their semi-steppe and desert holdings. Houses and stock shelters in these areas were allocated to individual cooperative members, and by early 1999 most had been dismantled and transported south to the vicinity of the main village for use as building materials. The cooperative also divested itself in 1997 of all its sheep, associated equipment and infrastructure, and employees that had been involved in its sheep operations. About 10,000 sheep and the buildings used to care for them were distributed in one massive allocation. Land titling was not a part of this exercise. Private sheep owners retained grazing rights to all pastures within the territory of the old collective farm, because – as they saw it – they had not withdrawn their land shares from the pool of common land held by the cooperative. As a consequence, private barns routinely sat on cooperative land.

The distribution of cooperative sheep holdings was both successful and generous to individual shepherds. The cooperative has, however, held tenaciously onto those classes of property that were essential to its operations or vital to the prestige and local authority of cooperative mangers. Arable land in the foothills was one of these critical resources. Use rights were generously granted to village residents who wanted to garden or to collect fodder; resources of this kind were abundant and their use did not challenge the authorities. Those who wished to register private title to land did challenge the cooperative's monopoly of land ownership and grain production and were treated very differently. Less than 100 ha have been registered under private title to a total of seven farmers who reported intentional administrative delays, high registration costs and persistent opposition by cooperative managers and ordinary members who viewed private land allocations as a threat to their interests.

The cooperative's defence of its dairy interests was equally effective. In 1999 the cooperative switched allegiance from an old state-run processing plant that never paid its debts and starting supplying milk to a commercial diary which installed a bulk milk storage tank in the village. The cooperative maintained a herd of about 350 milk cows, and could fulfil the village's effective milk quota, which was determined by the size of the storage tank. Because it was easier to deal with a single supplier, the dairy took cooperative milk in preference to that from village households, and paid the cooperative slightly more per litre than small producers. While the commercial dairy would in theory purchase milk from peasant producers, in fact most of them marketed their milk through private milk retailers.

Cattle were more profitable than keeping sheep in Kazakstan in the late 1990s, provided one could market the milk. Having sorted out dairy marketing, the cooperative had monopolised this lucrative business. In contrast to a Soviet collective, Foothill Farm had abandoned any pretence of involvement in all aspects of the village economy. By divesting itself of unprofitable enterprises, shedding peripheral assets and employees, and covertly opposing private producers who challenged its dominance, the cooperative was transforming itself into an aggressively managed, specialised, Western-style producer cooperative. But this was a cooperative with a difference; its directors – like Soviet farm managers – still controlled the land and dominated the community.

Withdrawal from marginal areas – Desert Farm

In 1965, the first year of L. I. Brezhnev's leadership in the Soviet Union, a law was passed to create in Kazakstan 150 new sheep *sovkhozes* with 50,000–60,000 head each. Between 1967 and 1974 in what is now Dhambul *raion*, Desert Farm (Table 5.1) and two adjacent state farms were created as part of this effort.

Desert Farm never supported the prescribed number of sheep because there was not enough pasture. According to local authorities and farm employees, the *sovkhoz* was also unprofitable from its creation, 'accumulating debt like a kind of subsidy', and it went into the privatisation process heavily in debt. At privatisation in 1995–6, the new cooperative assumed the bulk of the old *sovkhoz* debts – 22 million tenge – and paid these off by selling vehicles and animals. By early 1999 the cooperative had reduced its debt to two million tenge, but had lost most of its animals and equipment.

In 1996 soon after privatisation, large numbers of cooperative members claimed their shares and left the cooperative. This exodus created sixteen new agricultural enterprises and/or peasant farms. In the estimation of the village authorities, however, about half of these farms had existed only on paper and only two farms were still functioning by 1999.

There were multiple reasons for farm failures:

- Many of the seventy households that broke away from the cooperative were using the land registration process as a way of extracting their shares from the collective. As soon as shares were released, most withdrew from further cooperative work and either farmed by themselves or moved away.
- All the small farms formed in 1996 staked their futures on arable farming. Deprived of the economic distortions and subsidies that sustained it in the Soviet period, arable agriculture in the area was – at best – marginally viable. All of the private farms used their flocks primarily as a cash reserve to support

their arable activities. By 1999 none of these farms still used the stock shelters they were allocated because they had sold their animals to cover the running costs of other operations.

- Antagonism between the large cooperative and its smaller offspring contributed to the high rate of farm failure. Independent farmers at Desert Farm accuse the cooperative of unfair asset distribution (handing out sick animals, broken machinery or infertile land at inflated prices), overcharging for inputs or services, and either the late payment or non-payment of debts owed to individual farmers.
- The new farmers were also burdened with excessive registration costs for machinery and land, a topic that will be discussed in greater detail later in the analysis.

Facing huge practical and logistical problems as well as covert sabotage, the new and inexperienced private farmers probably did not stand a chance.

The new Desert cooperative chose much the same arable agricultural strategy as its erstwhile members and has declined for the same reasons – high production costs combined with low yields. With the exception of some fodder crops, cultivation around Desert Farm depended upon irrigation from a canal that had not been maintained for half a decade following the demise of the Soviet agency responsible for water delivery. Each year the canal delivered less water and the planted area declined. The cooperative has relinquished control over two-thirds of the area formerly owned by the *sovkhoz* (shrinking from 103,000 to about 35,000 ha), but retained possession of some land in all agro-ecological zones in which it formerly operated. It has thereby avoided taxation on land it could no longer use and retained a resource base consistent with a combination of irrigated and rain-fed cultivation and livestock husbandry – the diversified production mix of the old *sovkhoz*. This strategy was failing as flock sizes and cultivated areas inexorably declined.

An essential component of the reform process was a shift in economic priorities, from gross output to net profit. Farms like Desert *sovkhoz* were ill equipped to make this shift. They had been created to produce additional sheep and wool with little regard to the costs of meeting these objectives. In commercial terms, they were overbuilt, overstocked and overstaffed. Denied state support, many of these operations would have been uneconomic even in a buoyant commercial environment, and they stood little chance in the depression that accompanied economic reform in the 1990s.

Privatisation as asset stripping – Peri-urban Farm

If Desert Farm was an unavoidable casualty of reform, the same cannot be said for Peri-urban Farm. Peri-urban cooperative was formed in 1996 from one of the richest farms in Dhambul *raion*. Despite its large flocks and valuable arable land, by early 1999 the cooperative was facing legal action by creditors who were owed about 50 million tenge and was selling assets in order to remain in operation. As it collapsed, the cooperative had disposed of virtually its entire sheep flock without distributing a single animal to its members. It was also selling assets such as livestock shelters that on other farms were commonly retained for distribution to members. There was, however, one cooperative policy that favoured average farm members: senior farm managers countenanced private land titling.

It might appear that the farm was pursuing an aggressive free-market programme of asset sales and land redistribution. This was not the case. The privatisation policies pursued by cooperative managers can be better understood as a tactical response to the farm's unusual combination of a favourable location and financial collapse.

Directors and senior managers of failing farms have strong economic motives for prolonging a farm's existence. Destatisation reduced government's control over the salaries of senior farm managers who were now working for independent commercial enterprises. According to the testimony of disgruntled employees, many managers took this opportunity to award themselves inflated salaries. Subsequently, they made sure that these salaries were actually paid, even when their organisations were paying neither creditors' bills nor the salaries of ordinary employees. By selling collective assets to cover their salaries, senior managers could legally privatise the wealth of failing farms for as long as the farms could maintain a precarious existence and barter income could be used to circumvent sequestered bank accounts.

The imminent collapse of the farm also meant that these managers had little incentive to obstruct reform in order to retain control over land. In the case of Peri-urban Farm, cooperative members were also motivated to seek private land title. The cooperative owned valuable land that was close to town and could easily be subdivided and managed by individual households for small-scale commercial or subsistence gardening, orchards or grain cultivation. By the close of 1998, at least eighty-five private farms had been created and about half the rural land around the cooperative's central settlement was either privately owned or in the process of being registered.

Developments proceeded very differently on cooperative grazing land outside the central settlement area. In remote areas interest focused not on land but on the acquisition of the infrastructure that rendered it useful – stock shelters, houses and water points. Many cooperative shepherds planned to buy the houses and stock shelters that they had used as cooperative employees. Their only problem was scraping together the money or barter goods needed to make a reasonable offer. There were no restrictions on purchases by outsiders and those who had acquired barns had already obtained title to the land around the barn or were confident that they could do so.

There was, nonetheless, no rush to privatise pastureland comparable to the scramble for arable land around the farm's headquarters. Peri-urban Farm controlled seven separate land sections scattered in remote areas of the district and containing predominately low-value pastures suitable only for extensive livestock husbandry (see Figure 5.4). Only two of these seven sections contained any private plots in early 1999 and in both cases the proportion of the total area that was privatised was negligible.[1] Levels of private rangeland ownership therefore remained quite low even on a farm where other kinds of land were in high demand. There is a ready explanation for this state of affairs. Without domestic stock, extensive rangelands are unproductive and virtually worthless. Because of livestock losses since 1994, land tenure reform was taking place at a time when rangeland had lost its value because rural people had lost the capacity to exploit it. This meant that, irrespective of official policy, the reform process was on hold for grazing land, which constituted the vast majority of the land area of Dhambul *raion*.

Law, rural realities and the management of tenure reform

'In agrarian reform,' Pomfret has observed with respect to Uzbekistan, 'implementation is everything' (2000a: 278). Much the same could be said for Dhambul *raion*. Since about 1995 the problem has not been inadequate legal recognition of private rights to land but the erratic implementation of existing tenure laws. The overall direction of the land reform process has been influenced by national legislation, but the details of reform have been determined by economic conditions in particular communities refracted through the perceived interests of the local elites that control the reform process. In the absence of vigorous intervention by national agencies, the reform process has been characterised by variability and arbitrariness at the local level, what Pomfret has termed 'entrepreneurial corruption at lower levels of government' (2000a: 279).

The contrast between Foothill and Peri-urban Farms – both former *kolkhozes* located in similar agro-ecological areas – illustrates the problem. In Foothill Farm, cooperative members withdrawing from the collective received livestock, barns and houses as part of their severance packages. Peri-urban Farm never distributed livestock to its members except as salary payments, and required house and barn owners to purchase these structures, even if they were already living in them. In Peri-urban Farm barn owners expected as a matter of course to receive private title to the land around their farm; at Foothill Farm the opposite expectation prevailed. These contrasts were the result of economic differences rather than

---

1 For example, one northern section of 4000 ha had previously been used by cooperative flocks in the winter and early spring and had contained twenty-three winter quarters for sheep flocks. About eight of these twenty-three shelters – generally those with the best water supplies – were still occupied and in working condition in 1999. The rest had been demolished and the building materials carted away. Of the eight households still resident in the area, two worked as cooperative shepherds but expected to lose their jobs in the near future, two were 'employed' to guard the properties where they lived and kept private flocks, and the remaining four households combined various degrees of unemployment and self-employed shepherding. All these occupiers were interested in becoming owners but few had any real expectation of being able to raise the money to do so.

differences in legal status. Foothill Co-operative was financially viable because of its commercial dairy operation. While it could offer ex-members material benefits, managers were also interested in preserving the territorial integrity of the cooperative which was their power base. Peri-urban Farm, on the other hand, was bankrupt and facing complete dissolution. It had little to offer its members in economic incentives but its managers also had little to gain from thwarting aspirations for land ownership.

In the absence of mitigating circumstances, directors saw privatisation as a direct challenge to their authority and power. This did not foreclose other forms of support for private economic activity. Opposition by managers to private land tenure could co-exist with their granting generous terms under which cooperative members could privately use collective resources (as in Steppe Farm) or privatise property other than land (as in Foothill Farm). As a result of these attitudes, in Dhambul *raion* the privatisation of agricultural land has proceeded more slowly than any other form of rural property, despite Kazakstan's early reputation for radical 'fast track' reform policies.

Of all categories of land, remote pastures were the most unattractive candidates for privatisation. Pastures have remained overwhelmingly in the public or quasi-public domain either as state reserve land or as the property, often abandoned, of large cooperative farms. In the sample of farms covered in this study:

- Desert Farm has retreated to a core of land around the central village and abandoned outlying areas which were converted into state land.
- Unwilling to recognise formal change in the tenure system, Steppe Farm has quietly pulled back from its desert margins.
- Peri-urban Farm experienced rapid and thorough privatisation of productive arable land in the foothills versus almost no privatisation of its extensive rangeland holdings, despite a uniform approach to privatisation in both areas.
- Finally, Foothill Farm has abandoned its desert territories entirely and retreated to its more valuable holdings in the foothills, while the only attempt to privatise land in the cooperative was directed at this valuable foothill property.

The reasons for low rangeland privatisation rates are fairly obvious: semi-arid pastures were generally of low value per hectare at the best of times, and with the loss of Kazakstan's flocks they have become virtually worthless. The land's low value was in stark contrast to the high costs and complexities of the registration process. In theory, those registering land were obliged to pay fees that recovered the administrative costs of registration. These fees amounted to about 4,500 tenge according to notices posted in the district administrative headquarters and the estimates of land office staff. In fact, the costs associated with titling land, inclusive of transport expenses, were several multiples of this figure and in some cases up to 20–25 per cent of the value of the land being registered.

Faced with administrative opposition and high costs, poor and rich land claimants responded very differently. Wealthy and well-connected individuals were able to register ownership over resources such as shelters, water points and a symbolic amount of grazing land, and use the abandoned public domain for the bulk of their grazing. A small number of farsighted and successful producers were using this strategy to secure a diversified portfolio of holdings against the time when the pastoral industries recovered and grazing land again became a sought-after resource, a process described later in this chapter.

For poorer people the practical alternative to land registration was to remain within a collective agricultural enterprise. However, the security of these collectivised land shares remained unclear. It would appear that ownership of share certificates can be transferred; they can also be pooled and managed in common within a cooperative, the cooperative managers acting in this instance as trustees for individual share holders. The extent to which individual directors were managing other people's shares or had personally acquired these shares – possibly without the knowledge or understanding of the original owner – was unclear. Managers could also privatise collective resources by paying themselves high salaries at a time when other employees or outside debtors were being paid little or nothing. By the time the cooperative was formally bankrupt, managers may have conspired to transfer a considerable proportion of its resources into their own possession. Aside from fostering inequality, this approach to privatisation has prolonged the immobilisation of scarce resources under the control of defunct operations.

Recovery: apparent trends and unresolved issues

Because there is no scarcity of natural resources in the rangeland sector at the moment, because rangeland privatisation has hardly begun and because it has been so imperfectly implemented, the preceding case studies provide little empirical evidence on the optimal tenure system for pastoral Kazakstan. It is doubtful, however, that extreme fragmentation into numerous, precisely demarcated small-holdings – successful privatisation from the perspective of Kazakstan's current land tenure legislation – is the most effective way to manage an extensive, low-value, erratically productive resource like semi-arid rangeland. Evidence to substantiate this position is provided both by abundant comparative research from Inner Asia, semi-arid Africa and the ranching areas of North America, and by current developments in rural Kazakstan.

Nearly a decade after reforms began in Kazakstan, the general outlines of post-reform livestock husbandry and land use systems are becoming clearer. Commercial livestock businesses are routinely organised around extended families, frequently those that can draw upon non-pastoral financial resources. When flocks are small, these operations tend to be sedentary and village-based. However, large flocks are consistently more mobile than small ones and are managed from isolated farmsteads rather than villages. As recovery proceeds and private flocks grow in size, it is probable that household-based, migratory systems of stock husbandry will re-emerge.

The concluding section of this chapter discusses the kind of rangeland tenure system that will serve the needs of independent, mobile pastoral households. A large body of technical research and practical experience on pastoral tenure systems is briefly reviewed. This review identifies several reasons why policy makers in Kazakstan might want to avoid the parcelling-up of the nation's rangelands, and suggests ways this can be done.

The new entrepreneurs

In addition to changing the way large farms are organised and function, privatisation has created commercial agricultural production units of a kind unknown in the Soviet period. Much hinges on the future of these novel entrepreneurial operations, despite their relatively small size and marginal position relative to the large cooperatives. 'By now it is generally agreed that in every transition economy the growth of new private enterprises has been much more significant than the transfer of ownership of old enterprises' (Griffin 1999: 42). Creating favourable conditions for the success of these new businesses could, in the long run, more than compensate for policy mistakes made early in the reform process.

Thus far in Dhambul *raion* the growth of successful private pastoral operations has not been particularly impressive. Table 5.2 lists private flocks of over 100 adult animals in the areas of Desert, Steppe and Foothill Farms and one section of Peri-urban farm, covering a total of about 237,000 hectares. Table 5.2 is not a sample but an attempt to list all large flocks in this area.

One of the most striking things about the list in Table 5.2 is its brevity. It is probable that at least three or four flocks of around 100 head were erroneously excluded from this tabulation because their owners underrepresented their flock size, and a similar number of large flocks may simply have escaped notice. Taking these ambiguities into account, it is still likely that there were only twenty or thirty privately owned flocks of over 100 head in an area of well over 2,000 square kilometres. Successful private shepherding operations were thinly scattered and still rather rare.

The more successful of these operations possessed a characteristic social organisation based on kinship ties. With seven exceptions, all of the flocks in Table 5.2 were owned and managed by kinship units larger than a nuclear family. The successful family sheep business routinely included a father and his married sons, or the sons alone after the father's death, but collateral kin and relatives by marriage might also be included.[2] The exceptions to the general pattern were

---

2 In a separate survey of large flocks, it was found that three out of a sample of five large flocks were operated by extended families spanning several generations of adults (Chapter 8), a finding that is confirmed in Chapter 7.

Table 5.2 Large flocks in the study area

Reference No.	Location of home base	Flock size	Farm organisation	Land and buildings
1	Semi-desert	1000	Father and four sons	Land registered in 1994–5; barn
2	Semi-desert	700	Four brothers	Land registered 1998; barn
3	Semi-desert	400	Nuclear family and hired assistants	Barn; land under discussion
4	Foothills	350	Four brothers and sister	Barn, no land
5	Foothills	350	Father and brothers	None
6	Foothills	300	Nuclear	Barn materials purchased, no land
7	Semi-desert	230	Father and married son	Barn; land under discussion
8	Semi-steppe	200	Nuclear family	None
9	Semi-steppe	200	Nuclear family and hired assistants	None
10	Semi-desert	200	Kin by marriage	?
11	Foothills	150	Kin by marriage	One parcel of land in 1998 and another earlier; barn
12	Semi-desert	150	?	?
13	Foothills	135	Nuclear family	None
14	Semi-steppe	125	Father and sons	None
15	Semi-desert	100	Father and 5 sons	None
16	Semi-desert	100	Nuclear	None
17	Semi-desert	100	?	?
18	Semi-steppe	100	?	Barn, no land
19	Semi-steppe	100	Nuclear family and hired assistants	None

themselves interesting. In Table 5.2, flock no. 3 was owned by a recent Kazak migrant from Mongolia, and flock 8 by a Chechen family. Both of these families lived within a cluster of households of similar ethnic background and used shared ethnicity as a substitute for common kinship in building networks of cooperation and mutual assistance. Flocks 9 and 19, also owned and operated by nuclear families, represented the previous form of pastoral organisation based on wage employment rather than kinship. Both of these flocks were owned by shepherds who worked for a large cooperative. These private animals mingled with and benefited from inputs provided for the cooperative flock, including the labour of assistant shepherds hired by the cooperative.

Three flock owners in Table 5.2 held privately registered land. These were also the only families in the sample with access to lucrative non-pastoral sources of income. One flock (no. 11) was owned by an ex-director of Foothill Farm, one by a former stores manager for an agricultural research institute (no. 1), and the last (no. 2) was supported by the incomes from two brothers, one working for government in land registration and another in Customs and Immigration, both lucrative positions. Although the sample is small, there is a strong positive correlation between external sources of wealth, very large flock sizes and formal land ownership.

It would also appear that wealthy flock owners were not unduly inconvenienced by the normal restrictions on where individuals could claim land. Only one of the three pastoral land owners in the study area (no. 2) was situated within the *sovkhoz/kolkhoz* territory where the head of the household had been employed. The retired Foothill Farm director (no. 11) had been impatient to set up a private farm soon after independence and registered land on a neighbouring farm in the early 1990s. The ex-director had been born in the area where he claimed land, ran his private farm in conjunction with a brother-in-law who resided there, and had a brother who lived on and owned another private farm in the same area. When Foothill Farm finally countenanced land titling in 1998, the ex-director claimed additional land there on the basis of his prior employment. The former stores manager (no. 1) also received land on a neighbouring farm based, in part, on the claim that he was returning to ancestral clan territory.

In the course of this study, not a single purely arable private farming operation that was commercially successful was discovered. Such enterprises undoubtedly exist around urban areas where market gardening is an option, but commercial success took other forms in the more rural and pastoral areas that were the focus of this study. Aside from the predominately pastoral operations enumerated in Table 5.2, the most common, though not necessarily the most successful, commercial businesses were based on the ownership of a truck and a combination of haulage and small-scale livestock fattening, transport and trading activities. But the most successful non-pastoral agricultural businesses – and all those that held title to private land – followed no standard pattern. They instead exploited unusual commercial niches that suited their particular location and resources. One of these businesses was located in Desert Farm and consisted of a father and four adult sons who owned and repaired their own agricultural machinery consisting of a truck, four tractors, two combines and a flail chopper. They undertook contract work for local farmers and were often paid in animals that they fattened and sold. Equally idiosyncratic was a specialised fodder production business run by a former policeman in Foothill Farm, which catered to the fodder needs of urban and suburban stock owners around Almaty.

The guidelines that were supposed to specify where an individual could claim land did not rigorously apply to well-connected individuals, but there was little evidence of a land grab in the pastoral areas studied here. Successful private farming operations were commonly (i) built around a consortium of close relatives,

(ii) supported by wealth accumulated in a small number of lucrative positions (police, customs agents, accountants, cooperative directors, storekeepers, etc.). Few families met these two criteria and those that did had claimed so small an area of land as to constitute, as yet, little threat to the interests of ordinary rural residents and stock owners.

Changing patterns of pastoral land use

In Dhambul *raion* reform has precipitated a geographical realignment of people, animals and economic activity. Three mutually reinforcing trends are discernible – a withdrawal from isolated rural areas into villages and their immediate environs, a district-wide shift by both human and livestock populations back from the drier north to the better watered and more productive southern foothills, and a reduction in livestock mobility. Grazing pressure has been concentrated around villages and water points while the use of seasonal grazing resources, particularly high mountain or desert pastures remote from larger settlements, has declined sharply. Reform has, in short, encouraged the abandonment of extensive migratory pastoralism in favour of more settled forms of animal husbandry. What is not immediately clear is whether these changes are a transient effect of the reform process or a permanent feature of a market-oriented system of livestock husbandry. To resolve this question we must examine the new patterns of land use and the reasons producers cite for adopting them.

Evidence of the declining importance of the northern parts of the district is provided by a comparison of employment levels, property values and changes in human and livestock numbers across the four study farms and the district as a whole, as follows:

- Moving from north to south, Desert Farm has lost roughly 40 per cent of its population since independence; the population of Steppe Farm has marginally declined and that of Foothill farm has held steady. The population of Peri-urban Farm has probably grown, given the farm's prime agricultural land and its proximity to urban jobs and markets, but population figures for the farm are not separated from those for Uzunagach town, the *raion* administrative centre.

- Loss of formal employment within the village cooperative has also been heaviest for Desert Farm. In 1999 Desert cooperative employed 12 per cent of the workforce that it supported in 1995–6 when it was privatised, while Peri-urban, Foothill and Steppe cooperatives retained 30 to 40 per cent of their former employees.

- A similar pattern emerges with respect to the differential survival rate of sheep barns from north to south of the study area. Of 132 sheep shelters that existed at independence in the areas canvassed in this study, thirty-four barns or 26 per cent remained in operating condition by early 1999. Whether a barn escaped demolition depended in large measure on where it was located:

11 per cent (n = 66) survived demolition in the desert, 27 per cent (n = 51) in the semi-desert, and 86 per cent (n = 15) in semi-steppe and foothill areas. The further south a barn was located, the better its chances of survival.

• The propensity of sheep to survive the reform process also increased steadily on a north–south, rural–urban gradient, with the highest survival rates occurring in Peri-urban Farm, despite that cooperative's financial difficulties (Table 5.1). The generality of this pattern is confirmed in Table 5.3 which presents survival rates for sheep on twelve farms in Dhambul *raion*. Sheep survival rates for farms with their headquarters in the semi-steppe and foothills, 28 and 29 per cent respectively, were appreciably better than those for farms located in the semi-desert, at 18 per cent.

Table 5.3 Sheep survival and post-reform migratory patterns by major ecological divisions, Dhambul *raion*

Agro-eco zone	Farm no. (1)	Modal zonal flock size 1985–9 (2)	Total sheep 1998 (3)	Sheep survival rate (4)	Migration patterns – collective flocks
Semi-desert	1–3	81,100	14,500	18%	Last used mountain summer pastures in 1996 (No. 1) or 1994 (No. 2) but continue to use winter quarters in Sands. Farm 3 had no collective animals by 1999.
Semi-steppe	4–7	168,700	47,600	28%	Variable depending on location. Farm 4 last used mountain pastures in 1997 and continues to use winter quarters in Sands. Farms 5 and 6 both continue to use mountain pastures but either had no winter quarters in the Sands (No. 5) or kept only horses and camels there (No. 6).
Foothills	8–12	270,445	79,620	29%	Use mountain summer pastures but abandoned winter quarters in the Sands.

Notes
(1) See Figure 5.4 for the location of the central administrative villages for each farm.
(2) Including collective- or state-owned animals but excluding privately owned animals on sample farms.
(3) Including collective-, state- and privately owned animals on sample farms.
(4) Column 4 as percentage of column 3.

•   The private housing market reflected these population movements and shifts in economic activity. Within the district, houses of comparable quality increased in value from the semi-desert to the southern foothills. At the northern extreme in Desert Farm, abandoned houses had no resale value unless they were dismantled and the materials sold for use elsewhere, and whole residential blocks of the village had been demolished by their former owners. In contrast, in Steppe Farm some individual houses stood empty, especially two-storey buildings that were difficult to heat once the village-wide central heating system no longer operated. But there was no widespread dismantling of houses to retrieve building supplies. In Foothill Farm, as in other foothill villages, there was a market in residential properties, and families that moved out of the village could sell their old house without first tearing it down. Foothill Farm was, moreover, a destination for recycled building materials retrieved from places like Desert Farm and new construction was taking place.

The concentration of humans, livestock and economic activity in the southern third of the district has been accompanied by reductions in livestock mobility. By 1998 none of the large cooperatives in Dhambul *raion* followed what had in the late Soviet period been the most common migratory schedule for sheep – four main moves corresponding to the four seasons, winter quarters in the northern sands, spring and autumn on the semi-steppe or semi-desert, and summer in high mountain pastures. By 1999 cooperatives with their village centres significantly north of the rail line in the semi-desert retained their winter quarters in the sands and engaged in two major moves – to the sands in winter and to the semi-desert in spring, summer and fall. No collectively-owned animals from these cooperatives spent the summer of 1998 on mountain pastures. Conversely, cooperatives with their village centres just north of the rail line and south to the mountain foothills continued to use their mountain pastures in the summer of 1998. They also followed a truncated two-move migratory system: summer in the mountains and all other seasons in quarters in the semi-steppe or semi-desert (Table 5.3). These cooperatives had either abandoned winter quarters in the sands or never possessed such land.

Among private stock owners, seasonal stock movement was closely correlated with herd size, with larger flocks being more mobile than smaller ones. This pattern is also noted in Chapter 7. With the exception of one flock owned by a professional trucker, all small flocks (1–40 head) surveyed in this study were managed on a sedentary basis and the vast majority of these were managed from houses in villages. At the other size extreme, large flocks (100+ adults) were overwhelmingly mobile (making localised moves within a single ecological zone) or migratory (across major ecological zones) and none were operated from central village locations. Middle-sized flocks (41–99 head) were primarily sedentary but all were managed from rural or peripheral locations where grazing pressure was low. These results suggest that as flocks grow in size they will tend to become more mobile and spread more evenly over the landscape, the converse being true if flock sizes shrink.

Flock owners cite production costs as the reasons for these shifts in management practices since the initiation of reforms. Large flocks spread the costs of migration, which tend to be uniform irrespective of flock size, over more animals. By increasing the availability of natural forage, nomadism also helps to offset the expense of fodder provision, which otherwise remains constant per head (unless animal performance is allowed to suffer) and increases in absolute terms as flocks grow in size. The reverse holds true for small flocks, in which case it is more economical to purchase cultivated feed than to migrate in pursuit of free feed that is naturally available. Prevailing stocking rates also influence migratory decisions. Low over-all sheep numbers decrease the incentive to disperse by reducing competition between flocks for local forage; high stocking rates reverse the equation. These calculations suggest that there is nothing intrinsically uneconomic about migratory stock-keeping in a commercial setting. To the contrary, the recent shift to settled forms of husbandry has been dictated not by the adoption of commercial objectives but by the depressed economic conditions that accompanied their introduction – the massive destocking of the 1990s.

The history of pastoralism in Kazakstan supports this interpretation. Sedentarisation occurred once before in the last century, in the 1930s when Kazak pastoralists were forcibly settled in collectives formed around pre-existing foothill villages. Then as now, the effects of sudden impoverishment were compounded by the imposition of new property relationships and unfamiliar exchange and input supply systems. Although communism rather than capitalism was being promoted in this instance, the end results were much the same. Stock were eaten, hidden, expropriated or starved to death until there were so few that the remaining ones could survive on the pastures accessible from villages and settlements, and for a while migratory pastoralism ceased to exist. It took several decades of communist rule in the period between the early 1940s and late 1950s before livestock numbers rebounded sufficiently to require access to seasonal fodder supplies. During this period the economy had recovered enough to provide the investments in water development, transport and housing that were needed to re-colonise the northern deserts (Alimaev *et al.* 1986: 12).

Kazakstan's sheep population in 1998 was comparable to that of the mid-1940s. If history and economic incentives are a reliable guide to future developments, migratory pastoralism will recover in tandem with increases in flock sizes, stocking rates and rural prosperity. What is required – but may be difficult to deliver – are land tenure arrangements that facilitate this recovery. Both the means to implement these reforms and their content are open to debate.

Tenure policies for Kazakstan's extensive rangelands

Precisely because so little reform has actually taken place in rangeland areas, Kazakstan can still alter its approach to rangeland reform without causing undue disruption in the countryside. This is fortunate. Given the needs of family-based commercial pastoralism, a rethinking of Kazakstan's rangeland tenure policy is warranted in at least three areas, discussed below.

Titling

Based on the experience of transitional Asian economies, one observer has concluded that 'there are ... several ways to commercialise land and create an efficient land market without changing ownership titles' (Griffin 1999: 43). Decades of research on land tenure reform in developing economies supports Griffin's conclusions.

The availability of legal titles and cadastral surveys may be an important consideration for outside investors with the wherewithal to defend their interests in court. But for many peasants and herders, land is secured (or lost) through other means – by virtue of shared cultural values, by informal institutional arrangements or complex social ties at the community level. Small landowners require simple and cheap titling systems that reinforce and clarify these diffuse rights, rather than supplanting them. The current titling system in Kazakstan has probably had the opposite effect. Because it is complex, expensive and physically inaccessible to remote rural dwellers, the titling process has multiplied the opportunities for extortion by rent-seeking officials. It has also contributed to greater uncertainty and inequality of tenure rights. Uncertainty was promoted when titles were introduced, initiating a one-off struggle for long-term control of resources. Inequality has increased because politically marginal, poor or uninformed households have not, on the whole, been the winners of these contests.

Fragmentation

Following reform in the pastoral sector, kin-based herding operations have emerged, state farms have evolved into large cooperatives with uncertain commercial futures and minimal involvement in extensive livestock husbandry, and middle-sized commercial enterprises have failed to materialise. The eventual fragmentation of Kazakstan's rangelands into numerous, small private holdings is one possible outcome of current trends and policies.

The creation of private ranch properties has been successful in other sparsely settled pastoral areas of the world where it has been possible to create large holdings, often because European settlers excluded native residents from land ownership. Privatisation has had less to offer in areas where indigenous pastoralism has survived and large numbers of households must be accommodated on individual plots. Reduced and unequal access to resources is one problem. When the output of land per hectare is low and unpredictable, individual land allocations need to be large in order to safely sustain a household in both good and bad years. The creation of private holdings that are large enough to be economically viable implies the exclusion of many pastoralists from land ownership where rural population densities are comparatively high.

By forcing separate herds onto small parcels, fragmentation also diminishes the livestock carrying capacity and productivity of areas where forage availability is spatially and temporally variable (Behnke and Scoones 1993). Attaching flocks to small parcels of land restricts the ability of shepherds to avoid stress by moving away from localised problems such as drought or severe snowstorms. Fragmentation also deprives flocks of cheap and plentiful fodder supplies by interfering with their ability to harvest seasonally available forage production in different agro-ecological zones.

Finally, privatisation may have negative environmental impacts. Under semi-arid conditions, intensification on private holdings can increase agricultural output and improve resource conservation (Tiffen *et al.* 1995). But it is equally possible that poor landowners in a variable climate will be forced to forgo the theoretical advantages of private property by heavily discounting the future benefits of a restrained level of resource exploitation. The use of privatised rangeland for intensive fodder production or as a grazing reserve may also increase pressure on any remaining common grazing areas. In this case, over-exploitation occurs because private stock owners hold their animals as long as possible on common pastures, maintain more animals as a result of intensive forage production on private land, or reduce the size of the commonage by additional enclosure. Processes of this kind may have contributed to the best documented instance of land fragmentation and degradation in Inner Asia – the privatisation of the rangelands of Inner Mongolia (Williams 1997).

Co-management

For both industrialised and developing countries, it is accepted that different interests – private individuals, groups and the state – have legitimate land ownership rights. Reconciling the conflicting entitlements of these parties is termed co-management. Based on the experience of other pastoral regions, it is possible to anticipate the prospects and some of the likely problems for rangeland co-management in Kazakstan.

Communal range ownership and management is recognised in Mongolia, where pastoralists are politically, economically and demographically dominant. In parts of western China, where pastoralists are members of minority ethnic groups, rangeland co-management survives only because the authorities do not have the capacity to implement a programme of thoroughgoing privatisation (Banks 1999). Recently, the authorities in semi-arid Africa have reluctantly recognised co-management as a feature of pastoral land management. In Africa the co-management debate has focused on the legal status of customary law and of rural communities vis-à-vis civil law and the state. Few African states have been strong enough to completely substitute modern institutions of their own devising for the customary institutions that informally govern resource use within rural communities. The rise of non-government organisations, research-induced shifts in donor policy, and the financial rigours of structural adjustment in the 1990s further strengthened the explicit recognition of a degree of local autonomy. Legislation to this effect is currently under discussion in most African countries and could constitute the most radical shift in African tenure policy in half a century, although countervailing pressures exist and it is too early to judge the outcome (Hesse and Trench 2000; Toulmin and Quan 2000; Wily 2000).

Remarkably similar political dynamics have driven debates about resource management, government regulation and producer rights in industrial economies. In the twelve western states of America where range livestock production is important, the United States federal government owns about half of the land. Since the 1930s those parts of federal land used for grazing have been managed by a volatile combination of civil servants, local producer groups and politicians. Lacking 'indigenous' organisations comparable to Asian or African communities, American ranchers created lobby groups, built links to sympathetic politicians, and organised locally to defend their interests against regulatory agencies and competing claimants to federal grazing land. Until the 1980s when their position was undercut by fundamental shifts in the economies and demographic make-up of the western states, ranch interests were remarkably successful (Calef 1960; Foss 1960; Libecap 1981).

In none of the examples cited above did national authorities and political elites willingly compromise their power. Organised local interests challenged their hegemony overtly or covertly. In this respect Kazakstan is unfortunate. In Kazakstan collectivisation occurred early under Stalin, was ruthlessly executed, and subsequently reinforced by repeated attacks on private property and Kazak social organisation (*Central Asian Review* 1962; Channon and Channon 1990; Olcott 1995: 348). Farm reorganisation reinforced these trends. Early soviets were little more than the amalgamation of several villages or nomadic encampments under the leadership of a traditional local elite. In the following decades, however, the continuity between collective and pre-collective community organisation steadily eroded. Units of land ownership and agricultural production grew larger, collective farms (*kolkhoz*) in which workers were shareholders were replaced by state farms (*sovkhoz*) in which workers were government employees, and technicians replaced local elites as farm leaders (Olcott 1995). By the close of the Soviet period, household-based pastoralism and indigenous community organisation had been displaced in the countryside by vast, vertically integrated, mechanised farms staffed by an industrial workforce. When recent reforms undermined the viability of these monoliths, there existed no alternative social institutions equipped to fill the organisational void.

It will take time for former employees to develop local organisations independent from the old Soviet rural organisations and their governing elites. To a large degree, these institutions will need to be created – as they have been in industrial democracies – rather than simply rediscovered, following the more common Asian and African pattern of adapting and strengthening pre-existing indigenous organisations. Should successful collective resource management re-emerge in pastoral Kazakstan, it will be based on communal institutions very different from those that dominated in the Soviet period. This is not a process of transition or reform, but of fundamental institutional development. Both international and local precedents suggest that the process will receive little support from the rural authorities whose power it is likely to circumscribe.

As is evident from the preceding discussion, resource management in rangeland areas is not just a technical undertaking. In places like rural Kazakstan it is also a collective enterprise which draws upon the vitality and legitimacy of local producer groups. Like few other agricultural development endeavours, successful communal resource management depends upon and contributes directly to the realisation of one of the ultimate goals of the reform process – the creation of a vigorous civil society.

Conclusions

Kazakstan is the only ex-Soviet Central Asian republic in which the agricultural sector shrank dramatically in the 1990s, relative both to what it had been in the beginning of the decade (–18.9 per cent average annual decline) and in comparison with other sectors of the national economy (from 27 per cent of GDP in 1990 to 9 per cent in 1998) (see Table 2.1 on page 17). The crash in sheep numbers in the mid-1990s can be attributed to the loss of Soviet wool markets combined with the use of sheep as a substitute currency during periods of hyperinflation. Massive losses of animals in the state and large cooperative sector were not offset by an increase in private animals. The effects of these losses have been far reaching. In the pastoral sector, reform has destroyed capital in animals and livestock-related infrastructure, rendered inaccessible productive rangeland areas, and distorted the balance between settled and migratory forms of stock husbandry. In institutional terms, the reform process has left intact a rural elite with a vested interest in the status quo, a land reform programme that has barely begun, and national authorities with little apparent capacity to address these problems.

While suitable for arable areas, current legislation to privatise agricultural land is inappropriate for extensive, low-value, erratically productive rangeland areas used by numerous smallholders. The re-emergence of mobile, household-based pastoralism suggests the need for resource co-management by private owners, producer groups and government representatives. This approach has proved successful in other semi-arid rangeland areas. In Kazakstan, unlike many parts of pastoral Inner Asia, viable producer groups probably cannot be based on the revival of indigenous local organisations. The length and severity of collectivisation eliminated these social institutions and it may be necessary for rural communities to create modern, representative organisations to take their place. Tenure reform in rangeland areas may, therefore, be a protracted process, though one that will make an important contribution to meeting the overall goal of democratic reform in the countryside.

6

NEW PATTERNS OF LIVESTOCK MANAGEMENT

Constraints to productivity

Iain A. Wright, Nurlan I. Malmakov and Hélène Vidon

Introduction

Dissolution of the state farms, decontrol of markets, and other large-scale economic changes that swept through Kazakstan in the first half of the 1990s had very direct effects on the way livestock were managed. These changes have in turn affected the productivity of livestock, with implications for the people dependent on livestock as well as for the national output of this sector. The principal adjustments in management have been in the way livestock are fed over winter, the breeds that are kept, the seasonal movement of animals to different pastures, and the level of veterinary support. These changes are described in this chapter, which assesses the impact of the new feeding regimes and animal health status on sheep productivity. The chapter concludes with a simple model that highlights the impediments to sheep management following privatisation and offers some pointers for the future.

New forms of livestock management have emerged and many of the premises under which research on livestock production was carried out are no longer valid. The farm and flock structures, economic environment and social context under which livestock are now kept are totally different from the past. Thus the way in which information from research was previously applied to develop management systems is often no longer valid – the objectives of a private livestock farmer are not the same as those of a state farm and differentiated management systems are emerging. There is a need to understand the ways in which those who manage livestock are responding to these changes so that first, research and development programmes can be designed to achieve maximum effect and second, policies can be targeted to re-invigorate the livestock sector.

Methods of study

Information on past and current management and grazing practices was collected from the literature, from Kazak scientists and from interviews with staff of former state livestock farms (directors, livestock specialists, shepherds, veterinarians), large- and small-scale private livestock owners. Fieldwork was undertaken between February 1998 and October 1999, with visits to the study area made by the authors during at least one season each year. Information was collected using semi-structured interviews and on some topics where quantitative information was required, by direct questioning of livestock managers or direct measurement on animals. Further details of methodology are given, as appropriate, below.

Livestock management at the end of the Soviet period

In the late 1990s, prior to the break-up of the Soviet Union, the following was a typical management system in the district in which the current study was under-taken (Dhambul *raion*, Almaty *oblast*). Most state collective farms (*sovkhozes*) had separate winter, spring, summer and autumn pastures. Most sheep spent the winter in the desert sand dune areas to the south of Lake Balkhash. Shepherds moved to temporary winter quarters with large flocks of sheep. The young, replacement ewes were usually kept in separate flocks from the mature ewes. Sheep were shepherded while grazing during the day, typically moving five to ten km from the winter barn, and were kept enclosed, often under cover, in the barn at night. Only in severe weather did the sheep not graze and supplementary fodder, in the form of hay and concentrates, would then be provided, being delivered regularly from the farm centre by truck.

Most specialised sheep farms produced hay. For example in 1983, of 697 specialised sheep farms in Kazakstan, only 98 could not produce hay (I. I. Alimaev, pers. comm.). Hay was cut from natural or improved pasture. Where irrigation was not possible, drought-resistant varieties of *Agropyron* were sown as an improved species to provide hay in areas with as little as 200 mm of rainfall.

In spring the sheep were moved back to the steppe or semi-desert area and grazed with their young lambs until early summer. At that point some flocks were moved to summer mountain pastures which provided a high level of nutrition during the period of lactation. Families moved with the flocks during the summer months and lived in yurts temporarily. In autumn the sheep were moved back to the steppe to graze. The autumn grazing areas tended to be different from those used in spring. Further details of the seasonal movements are given in Chapters 3, 5 and 7.

Breeds of sheep

Sheep are the main species kept on the steppe, semi-desert and desert. Goats, camels, horses and more rarely cattle are also kept in the rangelands. In the Soviet era, livestock breeds and breeding were controlled by the state. There was, from the 1930s, a strong emphasis on developing breeds that were suitable for different ecological regions of the country, and research was carried out by animal breeding scientists. The state controlled which breed could be kept in each region and cross-breeding was regarded as undesirable as maintaining the genetic integrity of breeds was deemed to be important.

By the 1980s there were approximately fifteen breeds of sheep in Kazakstan, which could be classified into five main types, depending on their fleece charac-teristics (Table 6.1).

Table 6.2 shows the Ministry of Agriculture's plan for the distribution of breeds in the different *oblasts* in 1986. Some regions had only one breed of sheep, while others had several. The fine-woolled and semi-fine-woolled breeds were bred mainly from Merino sheep crossed with other breeds to improve wool production characteristics or to ensure suitability to the environmental conditions in different regions. A major goal of animal breeders was to increase volume of wool produced per sheep, with less emphasis on quality.

The local Kazak breed is a fat-tailed sheep (or more accurately a fat-rumped sheep). The Edilbay is an improved strain of these local Kazak sheep. The preference in Kazakstan is for meat with a high fat content and their fat-tailed breeds produce carcasses ideally suited to that demand. Karakul sheep are kept in southern regions of the country. A further description of Karakul sheep with reference to Turkmenistan is given in Chapter 11. The fleece characteristics of some of the most important sheep breeds are given in Table 6.3.

Current sheep management systems

The main livestock species in the study area is sheep and so the remainder of this chapter deals mainly with sheep management systems. Collapse of the state farms and severance of the link between research and production units, coupled with the new economic conditions, has led to changes in breed composition in flocks. Newly privatised farmers have different management objectives as well as financial concerns from the former state farms. In the study area, a sample of eighteen sheep flocks was surveyed in 1999. The composition of the breeds in each flock is shown in Table 6.4.

Table 6.1 Classification of sheep in Kazakstan according to fleece type

Type	Diameter of wool (μm)
Super-fine wooled	16
Fine wooled	21–2
Semi-fine wooled	26–8
Semi-rough wooled	30–4
Rough wooled (includes Karakul)	>35

Table 6.2 Ministry of Agriculture's plan for distribution of sheep breeds in different *oblasts*, 1986

Oblast	Fine wool	Semi-fine wool	Semi-rough wool	Rough wool	
				Fat-tailed	Karakul
Aktubinsk			Kazak Semi-Rough Wool	Kazak Rough Wool	
Almaty	Kazak Fine Wool Merino, Arhar Merino	Degeres, Kazak Butt Stock, Kazak Meat Wool		Kazak Rough Wool	Karakul
Atyrau				Edilbay	Karakul
Dhambul	South Kazak Merino	Kazak Meat Wool			Karakul
Zhezkasan		Degeres	Kazak Semi-Rough Wool	Kazak Rough Wool	
Karaganda			Kazak Semi-Rough Wool	Edilbay Kazak Rough Wool	
Kzyl Orda					Karakul
Kokshetau	North Kazak Merino	Kazak Butt Stock			
Kustanai	North Kazak Merino				
Taldy-kurgan	Kazak Fine Wool Merino	Degeress Merino		Kazak Rough Wool	
Turgay	Kazak Fine Wool Merino		Kazak Semi-Rough Wool	Kazak Rough Wool	
Uralsk		Kazak Akjaik Meat Wool		Edilbay	Karakul
Akmola	Soviet Merino	Kazak Butt Stock			
Shymkent	South Kazak Merino				Karakul

Source: ADB 1997.

The majority of flocks had some Kazak Fine Wool sheep, the breed that was kept by the former state and collective farms in the study area primarily for wool production. However many flocks also had some Kazak Fat Tail sheep, more suited to meat production. In some cases farmers were crossing their Kazak Fine Wool ewes with Kazak Fat Tail rams to produce a cross-bred animal. The reason farmers gave for the shift away from wool-producing breeds was the low price of wool and

Table 6.3 Wool production from some Kazak sheep breeds

Breed	Sex	Live weight (kg)	Fleece weight (greasy) (kg)	Staple length (cm)	Wool diameter (μm)
Kazak Merino	Ram	80	5.0	9.0	20.6–25.0
	Ewe	50	2.1	8.0	20.6–25.0
S. Kazakstan Merino	Ram	75	5.0	9.0	20.6–25.0
	Ewe	48	2.2	8.0	20.6–25.0
N. Kazakstan Merino	Ram	80	5.5	9.0	20.6–25.0
	Ewe	50	2.6	8.0	20.6–25.0
Semi-Fine Wool	Ram	84	3.7	13.0	27.1–37.0
(cross-bred)	Ewe	50	2.1	12.0	25.1–34.0
Semi-Rough Wool	Ram	80	2.5	n/a	n/a
	Ewe	53	2.2	n/a	n/a
Kazak Meat Wool	Ram	85	3.2	11.0	27.1–34.0
	Ewe	55	1.9	9.0	25.1–34.0

Source: Kazak State Agricultural University.

Table 6.4 Breed composition of a sample of sheep flocks, Dhambul *raion*, 1999

Breed	No. of flocks
Kazak Fine Wool	3
Kazak Fine Wool and Kazak Fat Tail	8
Kazak Fine Wool and Kazak Meat Wool	2
Kazak Prolific	1
Kazak Fine Wool, Kazak Meat Wool and Kazak Prolific	1
Kazak Fat Tail and Kazak Prolific	1
Kazak Fine Wool, Kazak Meat Wool and Kazak Prolific	1
Kazak Meat Wool	1
Total	**18**

the growing demand for meat. One scientist also commented: 'Farmers prefer meat breeds because they do not need a lot of fodder and they can graze more easily on semi-arid desert.'

While there have been great changes in the structure of the livestock sector and the situation is changing rapidly, Kerven (Chapter 8) has identified three main categories of livestock owner and management systems that are emerging:

- small-scale private livestock-owners
- large-scale private farmers
- cooperatives formed out of ex-collective farms

Small-scale private owners

Most households in the study area own a small flock of sheep (up to about forty) and sometimes one or two cows. Often the livestock do not represent the household's main source of income, but provide meat and milk for household consumption and a few animals for sale. However when the cooperatives were formed out of collective farms in 1994/5, many found that the number of people previously employed were too great to be supported by reduced farm resources. Many people left the cooperatives because they were not being paid and sought employment elsewhere. While it is common for one family member to be providing financial support from employment elsewhere (for example a son or daughter may be working in a city and sending money back to the household), there are many rural families with only their private livestock for support.

Among these small-scale farmers, flocks are kept indoors in household barns in winter (from late November or December until March, depending on the weather). In fine weather the sheep graze pastures near the village during the day. All sheep are usually fed hay, while some owners who can afford to, also feed concentrates and or/straw. The hay is either natural pasture which is cut in June, July and August or sown pasture, often *Agropyron* which was sown in the Soviet period as an improved species. Often hay is cut by hand over a three-month period, but individuals with access to machinery (a tractor and mower) can cut hay more quickly and can therefore make better quality hay in the early part of the summer before the vegetation senesces. Some individuals buy or barter for hay. In areas where cropping is practised by a cooperative, wheat or barley straw is often available, and is stored for animal feed in winter. Usually the cooperative does not charge for the straw. In some cases livestock owners buy or barter for concentrates, usually barley, and feed 200 to 500 g of barley per ewe per day. Sometimes concentrates are only fed immediately before and after lambing.

Lambing typically occurs in February or March, although some flocks start lambing in January. Sheep are normally turned out from barns in March (depending on the weather) and are shepherded around the village on natural pasture. Often small-scale livestock owners pool their animals for grazing during the day. They either pay a hired shepherd or take it in turn to shepherd the animals. The animals normally return to the village each night and therefore rarely graze more than five or ten kilometres from the village. In areas where cereals are grown the sheep graze on cereal stubble in autumn. Because of the poor state of repair of harvesting machinery, a considerable amount of grain is left in the field. One observer estimated that up to 20 per cent of grain may be left behind by the combine harvester. This grain is subsequently consumed by animals when they graze the stubble.

Occasionally sheep are sold from these flocks for money but often they provide only meat and wool for household use.

Large-scale private farmers

Private farmers are those whose main or only source of income is farming. They usually have more than sixty or seventy sheep and typically have 150–300 or even more. They are often technical staff from the former collectives with a sound knowledge of livestock management. It is common for these farmers to have obtained an area of land ten to sixty kilometres from villages, often based around a barn and a well which was used seasonally by the former collectives. These barns are rented or bought from the cooperatives, usually with some houses where the farm family lives (this arrangement is more fully described in Chapter 5). These farm units, being away from the village, have the advantage of having access to forage supplies not utilised by village-based stock. Although these private farmers typically have only 50–200 ha of land privately registered in their names, their animals graze on cooperative land, for which there is little competition at present because of the recent large decline in livestock numbers. Depending on the ecological zone in which the barn is situated, animals may be moved for one or more seasons. Animals from barns in the foothills may be moved to mountain pastures in summer, while those based in barns in the semi-desert area may be moved to other pastures in the same ecological zone for one or two seasons. Those close to the sand desert area may be moved for the winter to another barn in the sand dune area. Farmers based on the edge of the cropping zone may not move seasonally, as they can graze their animals on cereal stubble in autumn and even grow their own fodder crops for winter feeding.

Sheep are kept in the barns in winter with daytime grazing. Usually hay is fed and in some cases concentrates are also fed, especially in late pregnancy and after lambing.

In the early years following privatisation there was no evidence of any planned production system, and animals were sold or bartered when the household needed cash or goods. From 1998 there is some evidence of some private farmers beginning to modify their management practices and to plan their marketing (see Chapter 8). For example, one flock owner, with almost 300 ewes and 300 Angora does, had started to mate a proportion of his ewes in August so that he could sell lambs in the Almaty market three months earlier than his main crop of lambs, when prices were higher.

Most of these private farmers have expanded their flocks considerably since they were established, and some now employ shepherds. In other cases several male members of a family have pooled their stock to create a single large flock, with each relative contributing finance or labour (Chapter 8).

Cooperatives

As outlined above, prior to privatisation, each state or collective farm had land in different ecological zones, and livestock were moved to a different zone in each season. These farms had large numbers of sheep (25,000 to 65,000) with a well-planned management regime. As the numbers of livestock owned by cooperatives have fallen dramatically since privatisation, most cooperatives in the study area now have only 1,000 to 4,000 sheep and are in severe financial difficulties. Most cooperatives no longer follow the traditional seasonal migration, because the massive reduction in livestock numbers has reduced grazing pressure on the steppe and semi-desert (see Chapters 4 and 7). Another reason for the reduction in mobility is that the new cooperatives often no longer have the resources to move livestock. For example, they may not have the money to buy diesel to transport animals or the ability to make fodder to send to the traditional northern sand dune wintering areas. Only the largest cooperative in the area, with 21,000 sheep, still sent shepherds and sheep to the northern desert in 1998.

The highest mountain pastures are no longer used in summer because of the reduction in stock numbers, the lack of resources to transport animals and the poor state of repair of mountain tracks.

Most cooperatives now keep their sheep on the steppe and semi-desert all year round, moving them to different pastures within these ecozones at different times of year. The management and feeding systems are thus similar to those employed by private farmers.

Sheep performance sample study

During 1998 and 1999 eighteen sheep flocks owned by private farmers and cooperatives in the steppe and semi-desert study areas were visited and basic production data were collected. Visits were made on three occasions: in January (mid-winter) when nutritional status would be most critical, in April after lambing and in October before mating. The purpose of this exercise was to assess if there was any relationship between seasonal feeding practices and body condition of ewes in autumn at mating time and in spring after lambing. Nutritional status had been pinpointed as the most crucial limiting factor affecting sheep productivity under current management conditions, on the basis of informal interviews with a number of shepherds and livestock specialists conducted in 1998 in the study area. We hypothesised that ewes which had been grazed over summer in the mountain pastures would have received better nutrition prior to mating in autumn, and that higher nutrition levels in winter would be positively associated with better lambing rates in spring.

Body condition score was recorded on each occasion using the system of Russel, Doney and Gunn (1969). The amount of fat cover was assessed by palpation of the lumbar area and a score (0–5 scale) allocated to each animal. In flocks of less than 100 sheep, the body condition score of all ewes was recorded, while in flocks of over 100, a sample of 100 was selected at random. Body condition is a good indicator of the nutrition that the animal has received over the preceding few months. It should be pointed out that this system does not take account of the fat stored in the rump of the fat-tailed breeds, and so the fat reserves of these breeds will be greater than a non-fat-tailed breed at the same body condition score.

Table 6.5 Production data from eighteen flocks, Dhambul *raion*, 1998

	Mean	Standard deviation
Body condition score		
January (mid-winter)	2.18	0.317
April (after lambing)	1.36	0.136
October (mating time)	2.56	0.093
Lambing percentage (spring 1998)	119	24.7
Lamb mortality (%)	10.3	8.59

Nevertheless the technique is valuable for providing comparative data. Results of the body condition scoring are shown in Table 6.5.

The mean body condition score in autumn was 2.56 (Table 6.5). This is slightly lower than is deemed ideal for extensive hill sheep production systems in Britain, where it has been recommended that ewes should be at body condition score 3.0 at mating (Rhind 1995). The flocks which had spent the summer in the mountain pastures were no fatter in October than those that had spent the summer on the steppe and semi-desert, with mean body condition scores of 2.51 and 2.59 for the two groups respectively. In the Soviet era it would have been expected that sheep that spent the summer in the mountain pastures would have had higher levels of body condition in autumn (I. I. Alimaev, pers. comm.). The fall in the sheep population over the past ten years has led to higher biomass of vegetation in the steppe and semi-desert areas than ten years ago (see Chapter 4). This means that the sheep which have grazed mountain pastures in summer do not increase in body condition any more than those on the steppe and semi-desert.

Over winter all the sheep lost a considerable amount of body condition, with a mean body condition score of only 1.36 after lambing. To set this in context, it is recommended that for sheep systems in Britain, ewes below a body condition score of 2.0 before lambing should receive preferential treatment (Meat and Livestock Commission 1983). Clearly the level of nutrition over winter is low. Loss of live weight and body condition over winter is, in itself, not necessarily undesirable provided sheep are in reasonably good body condition in autumn, and this is expected in most sheep production systems. However the leaner the sheep are at lambing, the lower will be the lamb birth weight and the higher the potential lamb mortality. The mean lambing percentage (number of lambs born per hundred ewes mated) was 119 per cent (Table 6.5) but the standard deviation was 24.7 indicating considerable variation. At 10.3 per cent, the lamb mortality was reasonably low, but again there was considerable variation.

To explore further the relationship between winter nutrition and productivity, detailed information on winter feeding practices was collected from seven flocks. All these flocks had grazed near the home village during the previous summer. During winter all had been kept indoors at night and, weather permitting, had grazed around the village during the day. All sheep received hay in winter, although the exact quantity was not easily ascertained. Some flocks were also fed straw and/or concentrate. The hay was cut in June, July and August from natural pasture or sown pasture, mainly *Agropyron*. Hay had either been cut by hand over a three-month period, or, if individuals had access to machinery (a tractor and mower) cut mechanically, early in the summer before the vegetation had senesced. The hay cut by mower was invariably of better quality. Some farmers had bought hay from the cropping zone. Concentrates (usually barley), if fed, were purchased and sheep were fed approximately 200 to 500 g barley per head per day. Sometimes this was only done for a few weeks before and after lambing. The seven flocks were classified, according to the type of winter feeding regime, into two categories: fed hay plus concentrates (three flocks) and fed hay only (four flocks).

The mean number of lambs reared per ewe (to approximately four months of age) in each category is shown in Figure 6.1. The three flocks that were fed hay plus concentrates received about 200 g of concentrate per ewe per day and had a mean rearing number of 113 lambs reared per 100 ewes. This is in broad agreement with the lambing percentage of 119 found in the larger sample. In the four flocks that were fed only hay, the rearing percentage was 73. One of these flocks, with only eight ewes, had a high mortality rate (60 per cent). If this flock is excluded the mean rearing percentage was 86.

The better performing flocks tended to be the larger flocks. The mean flock sizes for the two categories in Figure 6.1 were 148 sheep for those fed concentrates and 18 sheep for those fed no concentrates. Interviews with large- and small-scale sheep-owners confirm that owners of small flocks lack the resources to be able to acquire concentrate feed and good quality hay.

Although the sample size is small and the results must be treated with caution, these data, coupled with the generally low level of body condition in spring, suggest that under the new sedentary farm of sheep husbandry, the level of winter feeding is a key determinant of flock productivity. The very low body condition of sheep in spring points to a general lack of winter fodder.

The trap in which small flock owners find themselves is typified by the situation of the worst performing flock in the sample. This small flock of only eight ewes,

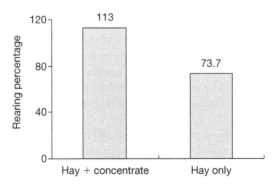

Figure 6.1 Winter feeding and rearing percentage.

which used to be larger, was on the edge of the cropping zone. The sheep grazed around the village on natural pasture all year. The owner collected some natural hay and straw from the stubble fields. He could not afford to buy concentrate for supplementation in winter. The sheep were in poor condition at lambing, lambs were weak at birth and mortality was high. Eight ewes produced five lambs in 1998, but three of them died. Performance of the flock was so low that there were not enough lambs being produced to maintain the numbers and the flock size has been progressively decreasing for the past four years.

Genetic potential

The genetic potential of an animal for a particular trait sets an upper limit to the level of production for that trait which can be achieved. Environmental factors such as exposure to disease, climate, level of nutrition and management then interact with the genetic potential to determine the actual level of production.

The traits of interest are determined by the objectives set by the manager of the system. If the primary product from the system is wool, then quantity and quality of wool produced will be traits of interest, while in a meat production system the number of lambs born per ewe per year and carcass weight and composition will be important. In practice, of course, there is usually a range of traits that are important in a successful production system. Animal breeders have devised methods for incorporating a number of traits into selection indices which are applied to populations of animals in order to select those which are genetically superior for a given production system. This latter point is important – genetic superiority for a trait can only be defined in terms of the objectives set for a system of animal production within a given set of environmental conditions. Take the case of a dairy cow that has been genetically selected for a very high milk yield in an environment in which feed supplies are abundant, placed in a situation where feed is scarce. The cow will produce milk at the expense of its own body condition, lose weight and become thin with a reduced probability of conceiving and a greater susceptibility to disease. In the long run, its lifetime output of milk may be lower than from a cow with a lower potential for milk yield, but which produces a calf regularly and lives longer.

Different breeds have different production characteristics. In fact different breeds have usually been developed by crossing and/or selection to produce the desired combination of traits suited for a particular management system in a particular environment. This is what lay behind the Kazak Ministry of Agriculture's plan of keeping different breeds for different regions, in the Soviet period (Table 6.2).

Different breeds of sheep produce wool of different characteristics (Table 6.3), with each type of wool suited for a different manufacturing purpose. Most of the fine and semi-fine wool-producing breeds in Kazakstan are based on Merino-type sheep. These have low levels of prolificacy (0.9–1.1). Historically in Kazakstan, pastoralists have selected fat-tailed sheep for meat production and survivability in severe conditions, with less attention given to prolificacy. Average prolificacy rates of 1.02 to 1.28 for fat-tailed sheep were reported by Ermekov and Golodnov (1976), Kanapin and Jumadillaev (1983) and Sadykulov (1985). The levels of prolificacy for the flocks of mainly Kazak Fine Wool and Kazak Fat Tail sheep in Figure 6.1 are similar to those reported in the literature.

Given the generally low levels of sheep body condition and the response in productivity to nutrition, especially winter nutrition (Figure 6.1), it would appear unlikely that the genetic potential of the sheep is limiting their reproductive rate.

Animal health

The financial difficulties experienced by state organisations in Kazakstan following independence from the Soviet Union likewise affected the veterinary service. Livestock production in the former state farms was supported by a national and farm-level veterinary service providing diagnosis, prophylaxis and treatment. The re-organisation and reduction of these services could be expected to have a major impact on livestock productivity under the new privatised conditions of livestock management.

Information on animal health and veterinary services in the study area was collected from July to October 1998 (Vidon 1999). Twenty-five interviews were held with veterinarians, staff from private drug companies and staff from the state veterinary services. A total of thirty-five employed shepherds or private farmers were questioned using structured interviews about the state of health of their flocks and their use of veterinary services and medicines.

Organisation of veterinary services

In the Soviet era there was a planned system of veterinary control under the Ministry of Agriculture. Now there is a self-controlled National Veterinary Committee, funded by both the Ministry of Agriculture and local administrations. The legal responsibilities of the National Veterinary Committee are first to develop and implement a programme to control infectious and parasitic diseases and second to conduct veterinary inspection of animal products. There are also eleven *oblast* Veterinary Committees on which depend the *raion* Veterinary Centres.

The Veterinary Committees are struggling to come to terms with the increase in the number of private farmers, making control of disease much more difficult than when livestock production was centralised under the state. The main state company which supplied vaccines has gone bankrupt, leaving vaccine production to small under-funded laboratories in research institutes. Vaccines are also imported, especially from Russia.

The National Veterinary Committee administers one-year regional plans against contagious diseases. The vaccination plan for sheep in Almaty *oblast* (including the study area) in 1998 was vaccination against salmonellosis, enterotoxaemia, anthrax, sheep pox and rabies, plus foot and mouth disease in Dhambul *raion* where there had been an outbreak in June 1998. In addition animals were to be tested for brucellosis and tuberculosis and treated for mange. There were thirty-six outbreaks of sheep pox in 1996, but following that outbreak a vaccination campaign was implemented and there was only one outbreak in 1997, suggesting that the mechanisms for controlling sheep pox have been successful in this case.

Generally vaccines are supposed to be provided free of charge. Veterinary surgeons charge for their services although in many cases veterinary surgeons employed by cooperatives provide a free service for members' private animals. Many animals are not vaccinated. Some private animals are not registered although livestock owners are supposed to register their livestock at the *raion* centre to aid in the planning of veterinary campaigns. Unregistered animals will not be vaccinated. Also a survey of five cooperatives in Dhambul *raion* showed that none of their sheep had been vaccinated for all the diseases included in the 1998 plan, partly because of shortage of vaccines.

Access to veterinary surgeons

Access to specialist veterinary services is relatively easy for cooperatives. All cooperatives still employ a veterinary surgeon and livestock specialists. Cooperative members can also have their private animals treated by the cooperative veterinarians. This service is free and only the veterinary medicines have to be paid for, although in some rare cases better-off cooperatives also pay for the medicines.

Private farmers sometimes call upon the services of the veterinarians at the nearest cooperative, but these veterinarians are less and less able to provide this service, as few cooperatives are on a sound financial footing. In many cooperatives the decrease in livestock numbers has meant a decrease in the number of veterinary surgeons.

Some *raion* veterinary stations have been given funding to employ veterinary surgeons to provide a service to private farmers, although the budget for this service is very low. There is great variation in how this system operates. In some areas farmers pay only for veterinary medicines, in others they also pay the veterinary surgeon for services. Although private veterinary clinics operate in cities, none have so far opened in rural areas, probably because veterinary surgeons do not consider these financially viable.

Access to veterinary medicines

Generally access to veterinary medicines is not too difficult, at least in Dhambul *raion* which is near Almaty, but many veterinary products are expensive in relation to what people can afford to pay. Supply of vaccines by the state is sometimes insufficient.

Cooperative members almost always obtain veterinary medicines for their private animals from the cooperative. Private livestock owners have much less access to the cooperatives. From interviews with private farmers, more than 50 per cent do not buy veterinary products from the neighbouring cooperative; three-quarters of these farmers buy them from the veterinarians of the *raion* veterinary stations. The others obtain their veterinary supplies from veterinary pharmacies, of which there are four in Almaty, two in the administrative centre of Dhambul *raion* and a few in other villages.

Prevalence of disease

The main infectious diseases should be controlled by vaccination (salmonellosis, enterotoxaemia, anthrax, sheep pox, foot and mouth disease and rabies). Brucellosis is either tested for or vaccinated against, while tests are carried out for tuberculosis. Although there have been some outbreaks of infectious diseases, generally the incidence of these diseases is low, although as already stated many private animals in small flocks are undoubtedly not vaccinated. Larger private farmers, who tend to be technically more proficient and aware of the risks associated with these infectious diseases, appear to be having their animals vaccinated. Also the massive reduction in livestock numbers, especially sheep, has probably reduced the risk of spreading infectious disease. But should livestock numbers increase again, there will be a risk of infectious livestock diseases spreading. There is a lack of coherence in the implementation of veterinary plans and poor coordination of vaccine provision. A major outbreak of disease would seriously overstretch the capability of the National Veterinary Committee to respond.

From interviews with shepherds there are four main non-infectious diseases in sheep in Dhambul *raion*. The most prevalent in terms of number of flocks affected is nasal bott, due to the sheep nostril fly, *Oestrus ovis*, which occurs in 60 per cent of flocks. The majority of affected animals are not treated but slaughtered for meat.

Sixty per cent of shepherds reported a coughing syndrome. In dry weather this is probably a reaction to dust inhalation, but in cold and wet weather is probably due to bacterial infection in the lungs. The number of sheep affected is usually low. Shepherds do not consider it a major problem and few treat it. Affected animals are slaughtered if they do not recover, although recovery is common in summer. Occasionally antibiotics are used but shepherds usually consider them too expensive.

Foot rot occurred in 55 per cent of flocks. It only occurs when sheep are kept in wet conditions – at the end of winter or on wet mountain pastures in summer. It is more prevalent in Merino-type sheep than the fat-tailed breeds. Treatment varies. In most cases affected animals are isolated and the necrotic tissue cut out. A range of topical treatments is then administered (petrol, formalin, etc.). Orf occurred in 46 per cent of flocks. Applying a local antiseptic, often petrol, usually treats this. Other diseases include scouring in lambs, ticks (especially in flocks that spend the winter in the sand dune areas and are not moved prior to the ticks becoming active), bloat (usually associated with eating particular plants) and gastrointestinal parasitism (11.5 per cent of flocks). Shepherds generally are unaware of the species of parasites causing problems. Some shepherds drench animals and two had used ivermectin.

Many shepherds reported using traditional remedies, including medicinal plants, for treating diseases and ailments, but a detailed study of these and their effectiveness was not possible. With an increasing interest in the use of medicinal plants for veterinary treatments in many parts of the world, such a study would be worthwhile to document local knowledge.

Although some health problems are widespread, their prevalence within a flock is generally low. The main conclusion of the study is that disease does not appear to be a major factor reducing productivity in sheep flocks.

On the basis of the information presented in this chapter and elsewhere in this book a simple model of flock performance has been developed (Figure 6.2). The output from a flock of sheep (lambs and wool) is used for three purposes: to provide meat and fibre for household consumption, to provide replacement animals to maintain the flock size and finally to provide animals for sale to obtain cash for basic necessities such as food and clothes, or investment. The output of the flock is a function of the performance (or output) of individual animals and the number of animals (flock size). Four key factors potentially influence individual animal performance: nutrition, genetic potential (set by the breed), health and the husbandry skills of the shepherd.

Seasonal nutrition is influenced by the type and state of the vegetation, since most animals graze pasture all year round, even in winter unless the weather is very severe. The land tenure system, and property rights in general, influences the pastures to which flocks have access. Winter nutrition is strongly influenced by the quantity and quality of fodder that is fed. Fodder is affected by the type and quality of the land from which it is made, which in turn will be affected by land tenure arrangements. Fodder quality is also influenced by the availability of reliable machinery.

Constraints to livestock productivity

The model in Figure 6.2 allows a systematic analysis of the current constraints acting on sheep production systems and identification of some of the means by which these constraints could be overcome. Of the four key factors influencing animal performance, it has been argued in this chapter that two of these – genetic potential and health status – are not major limiting factors in the study area. Husbandry skills are of course variable, but owners of large private flocks are generally well-skilled in sheep husbandry, although some small flock owners either lack the skills or do not have the resources to manage flocks properly. The major limiting factor on animal performance identified by this study is therefore level of nutrition.

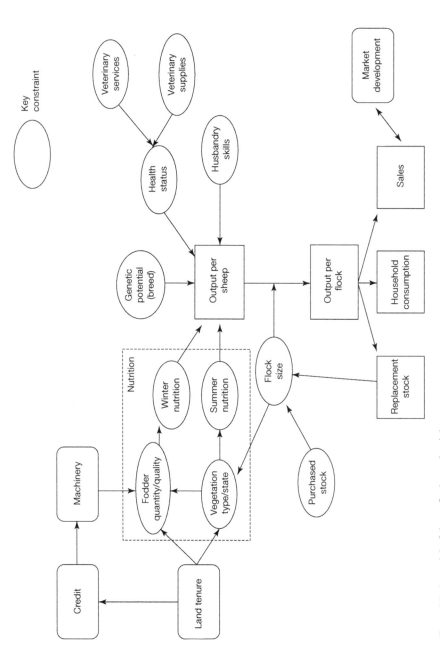

Figure 6.2 A model of sheep flock productivity.

Summer nutrition

Currently the level of summer nutrition is probably moderate, as indicated by the relatively low levels of body condition of sheep in autumn, especially those that graze on the steppe and semi-desert in summer. Ellis and Lee (Chapter 4) have suggested on the basis of remotely sensed data that there has been an increase in the biomass of the vegetation in the steppe and semi-desert areas as a consequence of the reduction in grazing pressure.

Winter nutrition

Undoubtedly, winter nutrition has a major impact on livestock productivity, as illustrated above. There is generally a shortage of winter fodder, and much of the hay cut from natural pasture is of low quality because it is often cut by hand over an extended period of time. Some areas are suitable for the growing of improved varieties of grass, e.g. *Agropyron*, but the spatial distribution of the current areas, i.e. very large blocks of land sown with improved species, is not appropriate to the needs of small-scale private farmers. There is a need to adapt the technology for growing these improved varieties to the needs of private farmers by ensuring a supply of seed and machinery suitable for their needs. This also requires that farmers have access to credit to allow them to invest in such technology (see below). Even access to simple mowing machinery would allow farmers to harvest natural hay more quickly in early summer and thus allow better quality fodder to be produced.

Access to pasture

While at the moment there is little restriction on where shepherds graze their sheep because the numbers are low, this will not be the case if sheep numbers increase in the future. Cooperative directors and large private farmers are beginning to control key resources such as wells and the better pastures, as documented in Chapter 5. Some of this control is official but much is not. This will lead to disputes about access to pasture unless appropriate land tenure arrangements are put in place. Land tenure systems must develop in a way that encourages rather than restricts the optimal use of the pastoral resources. Thus the option of seasonal use of different pasture types should be built into any land tenure system.

Flock size

Where livestock husbandry is not the main economic activity for a household, then small flocks can make a valuable contribution to the household food supply and income. Indeed the keeping of livestock is deeply rooted in the culture of Kazak people. However, there are many rural households with no employment and only a small flock of sheep, insufficient to sustain the household.

Table 6.6 Flock dynamics for family self-sufficiency

Sheep disposed of from flock	Per year
Household consumption	15
Sold/bartered	20
Die	3
Total disposed of from flock	**38**
Sheep added to flock	
Lambs born	38

The calculation in Table 6.6 shows that a family needs at least fifty ewes to form a sustainable flock, based on the following assumptions. With fifty ewes in the flock, about thirty-eight lambs will be produced each year, with a weaning percentage of 76 per cent. If it is assumed that a ewe will remain in the flock for five years and ewe mortality is 6 per cent per year, then thirteen ewe replacements will be needed each year to maintain flock size.

Thus theoretically a flock of about fifty sheep may maintain a family. It should be pointed out that selling twenty sheep per year would generate an income of only $600–700 per year. Thus the standard of living will be exceptionally low. A flock of less than fifty will not support a family as, after slaughter for household consumption and selling sheep to obtain other essential food and household goods, there will not be enough replacements to maintain the size of the flock.

There are many families that started in 1994/5 after decollectivisation of state farms with fewer than fifty sheep and by 1998/9 their flock size had fallen to less than twenty. These families live in desperate poverty and in the next couple of years they will have no animals left. Thus smaller flocks are often unviable if a household has little or no other income source.

Alternative employment

It is likely that these families, or at least the younger members, will move to the towns in search of employment. Indeed this is already happening. Whether alternative employment exists will depend on the rate at which other sectors of the economy develop. To avoid widespread migration, which could result in huge social problems, there is a need to encourage alternative forms of employment in rural areas. This could be based on downstream activities from the agriculture sector, e.g. small-scale food processing.

Investment and credit

There has been very little investment in rural areas for more than a decade and indeed much of the livestock production infrastructure in the form of water points, feed stores, winter shelters, and processing facilities has been destroyed. Agricultural machinery is in poor repair and farmers have difficulty acquiring the credit needed to make even small-scale investments. The provision of short-term credit would allow the opportunity for investment and expansion. Currently, apart from their animals, livestock farmers have no assets to act as collateral against loans. Land, instead of being an asset, is often a liability as an annual tax has to be paid. If land leases were long-term, tradable (which currently they are not) and heritable then this would provide security of tenure, which would in turn give leases the value needed for them to be offered as collateral. If this was coupled with credit schemes targeted at farmers, then much-needed investment would flow back into rural areas and allow efficient farmers to develop their businesses.

Marketing

The collapse of the state-organised marketing structures was initially a major brake on the redevelopment of the livestock sector. While new marketing channels are developing, they are currently weak in areas more distant from major markets and are moreover concentrated on meat offtake, leaving producers with very little returns on wool which was formerly a major economic output. From small flocks, returns from marketing are not sufficient for investing in flock improvement. An analysis of the constraints to livestock productivity imposed by the current marketing systems is given in Chapter 8.

Conclusions

There is undoubtedly much scope for development of the livestock sector in Kazakstan. The country has vast natural resources suited to extensive systems of livestock production. However the futures for the three types of livestock owner identified by Kerven in Chapter 8 are likely to be very different.

Livestock keeping is likely to continue to be important for most rural households. Where the household has an alternative source of income, a small private flock will provide a valuable source of meat for household consumption and contribute to food security for that household. Where the household has no alternative source of income and only a small flock of sheep of less than fifty ewes, then the future looks bleak. Those people will not be able to maintain their flock sizes and will be forced to seek alternative employment, inevitably accelerating the drift of population to urban areas.

It is difficult to see how the cooperatives that are remnants of the former state farms can survive in their present state. The few that still receive state aid in the form of credit and other inputs are operating relatively effectively. Others that supported large workforces in the past now find that without the massive state subsidies they used to receive they can only employ about half of their previous workforce. While many cooperative still have a considerable pool of technical expertise, most of them do not have the management skills to cope with a market-oriented economy. There is a chronic lack of capital investment and much of the state farm infrastructure has been destroyed.

The new larger-scale private farmers usually have sufficient resources, consisting of a sufficiently large flock, buildings, machinery, etc., to develop commercially viable businesses. They mostly have the technical managerial skills and are developing some of the financial and economic skills to operate in a free market. They do, however, represent only a tiny proportion of the rural population.

For the livestock sector to develop there is a need for policies to assist in technical innovation as well as economic and institutional reform. Two key technological and policy issues need to be addressed. First, is the development of appropriate technologies for the provision of winter fodder. Second, is the development of a land tenure system that promotes sustainable use of pasture resources for livestock production and provides security of tenure as a basis for investment in rural areas. There is also a need to develop rural credit schemes to finance both short-term credit and longer-term credit for capital investment.

7

CONTRACTION IN LIVESTOCK MOBILITY RESULTING FROM STATE FARM REORGANISATION

Sarah Robinson and E.J. Milner-Gulland

Introduction

A nomadic way of life has long been a defining characteristic of the Kazak people, both in the imagination of foreigners and in Kazak literature celebrating the steppe and nomadic life. Even today it features strongly in government efforts to create a national identity. However for over 150 years nomadism has been on the decline in Kazakstan, so that today it involves only a tiny proportion of the population. In this chapter the major factors which have affected the mobility of livestock in the twentieth century are reviewed, and changes occurring since independence in 1991 are examined in detail. A case study of changes in migration patterns is used to illustrate the points made.

The study area

The study on which this chapter is based was conducted in 1997 and 1998 in central Kazakstan.[1] The area includes central Dzhezkazgan *oblast* (province), plus the northern *raions* (districts) of Dhambul and South Kazakstan *oblasts* (see Figure 7.1).

The region was chosen mainly because it includes one complete migratory system which was in existence both before 1917 (see also Figure 7.2), and in a shortened form in Soviet times. Prior to the Soviet period, stock from what are now Dhambul and South Kazakstan *oblasts* wintered in the Moiynkum sand desert, and spent summer in Karaganda *oblast*, or even further north, pastures known as Sary Arka. Spring and autumn were spent crossing the clay desert of Betpak-dala,

1   Research for this study was carried out with funding from the EU INTAS programme for a project on 'Land degradation and agricultural change on the rangelands of Kazakstan'.

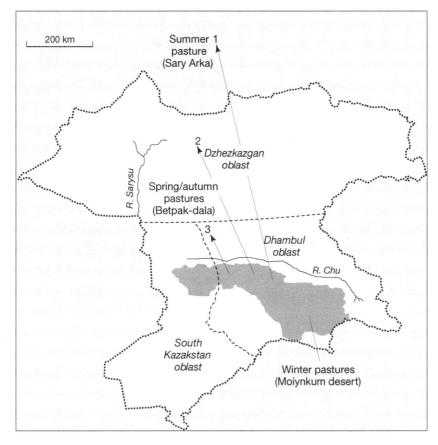

Figure 7.1 A map of the study area showing the Moiynkum–Sary Arka migration in the pre-Russian era (1), the later Soviet era (2), and the late 1990s (3).

and along the river Chu. Figure 7.2 shows the migration before 1917, in the late Soviet period, and in the late 1990s. This migration system provides a case study for the changes described in the next sections.

The ecological background

In the non-mountainous areas of Kazakstan there are three main ecological zones described by Soviet botanists (for example Kirichenko 1980 and Zhambakin 1995), in which differences in rainfall, snow cover, and vegetation type have all been important for the history of livestock mobility in Kazakstan. These zones are shown in Figure 7.3.

The steppe zone in the northern regions of the country has an average annual rainfall of above 300 mm per year, the greater part of which occurs in the summer,

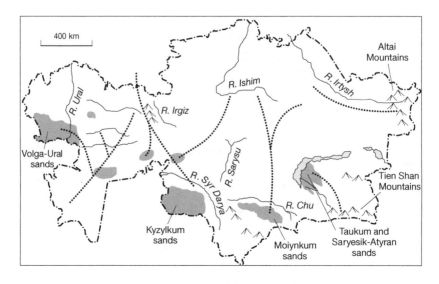

Figure 7.2 Some of the major patterns of migration of Kazaks in the fifteenth century.
Source: Zhambakin 1995.

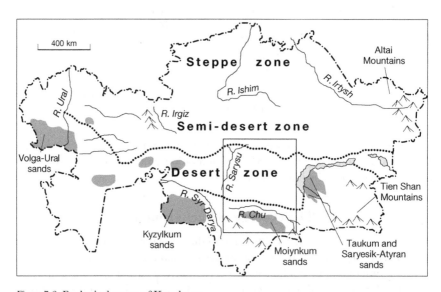

Figure 7.3 Ecological zones of Kazakstan.

Notes
Stippled regions show pastures on sandy desert soils. The study area is marked by a rectangle. The thick line represents the border of the clay desert zone.

making it ideal pasture at this time of year. The natural vegetation consists of pastures dominated by grasses such as *Stipa, Fescue*, and *Agropyron* species. The semi-desert zone typically has a rainfall of 200–50 mm per year, and is dominated by a mixture of grasses and *Artemesia* species.

The desert zone comprises roughly those areas having less than 200 mm of rainfall per year. This is the area southwards from about 47 degrees latitude. The northern desert zone is characterised by flat land on clay soils, and an almost total absence of grass species. Here productivity is low, and vegetation is dominated by species such as *Artemesia terrae-albae* and *Artemesia turanica* together with salt-tolerant and xerophytic species such as *Anabasis salsa, Salsola orientalis, Salsola arbusciliformis, Atriplex cana*, and *Kochia prostrata* which become increasingly dominant towards to the south (Kirichenko 1980).

South of the rivers Chu and Syr Darya, and Lake Balkhash, are the three largest sand deserts of Kazakstan, the Moiynkum, Kzyzlkum, and Taukum deserts (see Figure 7.3). These are characterised by shrubby vegetation such as *Haloxylon* and *Calligonum* species growing on dunes.

Snow cover in the semi-desert zone exists between November and March and has an average depth of 25–35 cm. In the steppe zone it is higher. Animals cannot obtain food under snow when the depth averages 35–40 cm, or 20 cm when the snow is dense (Sludskii 1963). Therefore, the steppe and semi-desert regions cannot be used as winter pastures if supplementary fodder is not available. The best winter pasture areas are on sandy soils in the desert zone due to their shrubby vegetation which can be browsed even under conditions of deep snow.

Kazak pastoral migrations

The ecosystems of Kazakstan are generally harsh, with low rainfall, and crucially, the effective elimination of a large part of the pasture in winter due to snow cover. The distances separating winter and summer pastures were often large, leading to long migrations of up to 700 km in some cases. The spring and autumn pastures were generally the northern clay deserts (*Artemesia* and saltwort pastures) lying between the summer and winter areas. In mountainous areas the foothills were used in spring and autumn.

Due to the constraints of long migrations, the Kazaks were purely pastoralists before Russia took an interest in the country, practising little agriculture or hay production at all. They thus did not have any supplementary feed for their animals (Matley 1994), and livestock numbers were entirely influenced by climate and seasonal fluctuations in pasture availability.

Particularly disastrous were *dhzuts*. This is a Central Asian term referring to conditions under which melting snow re-freezes to form an icy layer covering the grass, or to unusually heavy snow falls (Zhambakin 1995; Fadeev and Sludskii 1982). Such events could cause entire herds to die of starvation. For example, in the winter of 1879–80 in Turgai *oblast* over 40 per cent of stock died due to a *dhzut*, and a decade later up to 95 per cent died in the Kazalinsk area (Zhambakin 1995).

There is also some evidence to suggest that livestock populations were affected by drought. In 1928 for example one quarter of the livestock in south and west of Kazakstan died due to drought (Channon and Channon 1990). For Kazaks living in an extreme continental environment, migration was the best way of reducing the risks of *dzhut* and drought to a minimum, whilst at the same time maintaining high numbers of livestock.

The major migration routes were partitioned between three major tribes, or 'hordes' known as the Great, Little, and Middle hordes, and some of the major movements are shown in Figure 7.2. The migrations were not fixed, however, and depended from year to year on climatic conditions. For example if there was drought in the summer territory of one horde, it might move further north, or into the territory of another horde (Zhambakin 1995).

During the nineteenth century, most of the area of Kazakstan came under Russian domination. As outlined in Chapter 3, the Kazaks lost most of their best summer pastures, and migrations declined both in distance and frequency. In Aktiubinsk *oblast*, at the end of the nineteenth century, Kazaks were already moving only 20–40 km from their winter pastures, and in South Kazakstan *oblast* the majority of the population were sedentarised by 1908 (Zhambakin 1995). According to Olcott (1995), by 1917 almost 75 per cent of the population had some form of fixed winter quarters where many would also grow grain to feed their stock, and 50 per cent migrated only from May to September. In other words, they had become semi-nomadic.

In summary, up to the era of collectivisation there was already a great decrease in mobility among Kazaks, due mainly to diminished access to land. However nomadism did not disappear and migrations which did not enter agricultural areas, such as the Moiynkum–Sary Arka migration, continued.

After 1930 settlement was deliberately enforced upon the population in the form of collectivisation. The failure to sedentarise nomads was officially recognised in the 1940s and the system was adapted to a limited form of nomadism which continued within the context of the state farms until the early 1990s.

The Communist Party resolution of 1942 to increase the numbers of stock included provision for use of remote pastures. This occurred mainly from farms situated in the desert regions, whose animals started to migrate each year to the northern semi-desert or steppe regions with their rich summer pastures (Alimaev *et al.*1986). In the study area at this time, the migration from the Moiynkum desert went as far north as Karaganda *oblast* (Zhambakin 1995).

The increase in production targets required that new pastures be brought into use. However in the 1950s most pastures in the steppe region were ploughed up for grain production in the massive Virgin Lands Campaign, and any expansion of the livestock sector had to occur in the other vegetation zones. To this effect, in the 1960s, specialised sheep raising *sovkhozes* were created on the state land reserve in the semi-desert and desert regions, with a stock of 50,000–60,000 sheep each (Asanov and Alimaev 1990). This included new farms in the central region where this study was conducted. Wells were sunk, water was extracted by pump where the water table was deep, or even brought in by tanker, so opening up new areas for grazing.

Under the Soviet system, the farm structure in Kazakstan was predominantly that of the *sovkhoz* (state farm). These were state enterprises in which each worker had a wage paid by the state, with bonuses if the state quota was exceeded. Each contained in the 1980s an average of 540 workers and their families (Goskomstat 1988) and had a school, hospital, social facilities, veterinarians, accountants, technicians, shepherds, tractor drivers, and agricultural experts. *Sovkhozes* usually had defined boundaries which were marked on maps, and beyond which stock movement did not occur.

Despite the contraction in movement during the Soviet and pre-Soviet period, long migrations still existed in the study area. Although primarily conducted on horseback, migrations benefited from considerable technical support. Families could use tractors to help transport their yurts and food, and were supplied with fuel to keep the water pumps going. Veterinarians would visit the shepherds during the summer months, and in certain areas trucks would visit the shepherd and his family once a month with supplies of food. Those members of the shepherds' family who had to remain at the *sovkhoz* while the shepherds were absent (e.g. children of school age) were fed by the *sovkhoz* during the summer. Some migrations crossed the boundaries of the Kazak republic, spending one or two seasons in what are now Kyrgyzstan and Uzbekistan. Through these efforts the Soviets managed to make use of even the remotest pastures, and in the 1980s livestock numbers managed to reach a high of 36 million sheep, double the number in 1916 (see Figure 5.1 on page 76).

Two types of migration

Stock movement patterns in the Soviet period can be divided into two major types:

1   Long distance migrations across ecological zones. These usually involved movements to pastures on state reserve land remote from the *sovkhoz*, in other districts, provinces or even other republics. These migrations were several hundred kilometres or more in length.

2   Short distance migrations, which should probably be termed movements, occurring within the farm territory and which did not always entail a move with each season. These movements would have been between 10 and 120 km in length and were usually within one ecological zone.

The long distance migrations were mainly from farms in the south situated between a good winter site such as a sandy desert and a good summer pasture such as Sary Arka or the Ala Tau (Tien Shan) mountains. They would generally involve mass movements of hundreds of thousands of animals from entire districts.

The short distance migrations occurred mainly on farms further north in the semi-desert zone, in which each individual *sovkhoz* would be split into units of seasonal land use. On some of the new farms built in the 1960s, stock could be on the same pasture for three seasons, autumn–winter–spring, or spring–summer–autumn, with movement often constituting simply of rotations between adjacent wells. According to Zhambakin (1995), such grazing regimes lead to pasture degradation.

To take the example of Dzhezkazgan *oblast*, half of its area was remote pasture for animals from Kyzyl Orda, South Kazakstan, and Dhambul *oblasts* engaged in long distance migrations. The rest of the *oblast* area consisted of *sovkhozes* established in the 1960s, on which livestock would have been engaged in short distance migrations.

The difference between the two types of movement was never clear cut. Even in the more northerly *oblasts*, if significant amounts of particularly useful pastures existed, such as those on sands or along rivers, these would be set aside as state reserve land and used by several farms from the area which would drive animals there on a seasonal basis.

An illustration of a long distance migration (or short distance migration on a large farm) is described below, season by season.

Winter (October–April)

Shepherds lived in winter houses (*zimovki*) during this time. A *zimovka* is generally a brick construction, and has a barn next to it where the animals are kept at night and in bad weather. The *zimovki* are usually single isolated houses, but sometimes they are clustered together in small hamlets called *otdelenie*.

Spring (April–June)

Stock were moved from the *zimovka* to the lambing area for one month. This area was usually very close to the *zimovka*, on the edge of the winter pasture. This way the sheep did not have to migrate a long distance when pregnant, but could benefit from fresh pasture and also pasture which was never used for long, and was probably free of parasites. The sheep were sometimes moved once more in mid May to another spring pasture where they were shorn.

Summer (mid June–September, or May–September)

This pasture was generally the furthest from the winter pasture, and all shepherds lived in yurts or wagons at this time. However, in some farms in the semi-desert zone, summer pastures were just 10 km away from winter pastures. At summer pasture, each shepherd and herd were sometimes allocated two or more wells and a mid season move to a new water point was common.

Autumn (September–October)

This was sometimes in the same area as the spring pasture, and was often in the vicinity of the farm centres, lying between the summer and winter pastures.

To summarise, during the Soviet period and especially after the 1960s, the whole of Kazakstan's rangelands were partitioned into a patchwork of units, each with its land use and type defined by the state. Each farm, and indeed each flock and herd, was allocated grazing locations for each season. Shepherds took no decisions at all as to where they went with the animals in their charge. The farm managers took all decisions. However, upon closer inspection, it can be seen that some of the longer migrations which remained followed routes of earlier movements from pre-Russian times. For example, the migration from the Moiynkum desert to Sary Arka, and from the winter pastures of the Kyzylkum desert and Syr Darya river to summer pastures further north (see Figure 7.2) survived in a shortened form. In both cases stock did not go so far north, as the former summer pastures were taken up by new farms and stock which stayed all year round in those areas, barring in-migrations from southern farms. From the 1960s onwards animals went only as far as pasture in Dzhezkazgan *oblast* (still referred to as Sary Arka).

Effects of reform on migration

The major result of de-collectivisation was a huge drop in stock numbers (see Chapter 5). The loss of animals occurred both for collectives and new individual private farmers. In some cases animal numbers dropped due to starvation. Livestock *sovkhozes* were essentially in a semi-nomadic system totally reliant on winter feed, often produced elsewhere, which was paid for by the state. With the end of state subsidies on feed, mass mortality as a result of heavy snow occurred. For example, two farms in Karaganda and South Kazakstan *oblasts*, visited during the study, lost 5,000 and 30,000 sheep respectively in a matter of weeks.

With few animals, long distance migrations were no longer necessary or viable. The economies of scale needed to support the migrations no longer existed, all long distance movement in the country virtually ceased, and vast areas of Kazakstan were emptied of livestock.

The land tenure reforms accompanying de-collectivisation involved distribution of land shares by *sovkhozes* to their workers, as discussed in Chapter 5. However, these changes did not affect the remote pastures in the state reserve land as they were never *sovkhoz* property. Officially shepherds or the new collectives should pay rent for use of land in these areas. However, in 1998, the few people still going to the state reserve land in Sary Arka, which had formerly been designated for the use of their *sovkhozes*, did not encounter any obstacles to free movement.

Where several seasonal pastures existed on *sovkhoz* territory, land shares were offered to ex-workers as separate parcels, each in the different seasonal pasture areas. Therefore, continued migration was envisaged by policy makers. This is also suggested in the Civil Code of 1995, article 80 (Republic of Kazakstan 1995), which stipulates that *oblast* and *raion* authorities can grant permanent land plots set aside for the driving of stock to and from summer pastures, or arrange temporary tracks in agreement with land owners. However, the implication of this legislation is as yet far from clear. This is because so far access to land has not been a constraint to stock mobility, as is discussed below.

The impacts of reform in central Kazakstan

The effects of reform on long distance migrations is illustrated here by the migration from the Moiynkum desert to Sary Arka (Figure 7.1). Overall, of seventy-one farms in Dzhambyl *oblast* seasonally using land in Sary Arka in the 1980s, it was found that shepherds from only five farms continued to migrate after the reforms (with greatly reduced numbers of stock). None of these shepherds were independent private farmers, and they were all still employed by collectives.

Due to the loss of livestock, the numbers of sheep going to Sary Arka from the two northernmost *raions* of Dzhambyl fell from about 600,000 to less than 50,000 between 1992 and 1997. In 1997 the animals were spending winter in Moiynkum desert much as before, but were spending summer at wells on the former spring–autumn pasture on *sovkhoz* property.

This meant that all the state reserve land in Dzhezkazgan *oblast* which had been designated as summer and autumn pasture for other *oblasts* was empty at the time of the study (1997–8). The same applied also to the summer pastures in the Ala Tau mountains (Kerven *et al.* 1998), and it is probable that the same story is repeated throughout the country.

The effects of reform on short distance migrations in the semi-desert zone can be illustrated by the case of farms in northern Dzhezkazgan *oblast*. Here, at the time of the study the *sovkhozes* had been fully de-collectivised, and so the only farms in existence were private households. Therefore the behaviour of individuals who were essentially making all their own decisions could be assessed.

Of the private farmers who had been allocated land plots in different seasonal pastures, many were not using this land at all, but were pasturing their animals on common land around the village, or were just using winter pasture all year round. The frequency and distance of migration appeared to be directly connected to the number of animals owned by the farmer (Figure 7.4). However most people had received too few animals to make any movement at all worth their while. The case study below illustrates this point.

Sovkhoz M had 60,000 sheep in the 1980s. At the time of privatisation in 1992, assets were divided on paper, but were not given out until 1995 when fifty private farms were formed. In this year the *sovkhoz* had 19,000 sheep, 462 horses, and twenty cows left, which were distributed to all *sovkhoz* members, while other assets were only distributed to those starting a private farm. By 1998 more than twenty of the private farms on the former *sovkhoz* had folded and only 12,000 sheep remained on the farm. The frequency distribution of ownership among registered private farmers in 1998 on the farm is shown in Figure 7.4. This shows clearly that the largest category of shepherd is that owning fifty sheep or less; however this

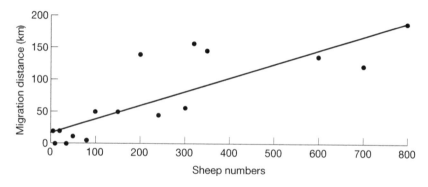

Figure 7.4 The relationship between sheep numbers owned by private farms and their migration distance between winter and summer pastures for a former *sovkhoz* in Dzhezkazgan *oblast* ($R^2=0.7$, P<0.001), 1998.

distribution concerns only the minority who had registered as private. The rest of the farm population tended to have even fewer animals.

This distribution highlights a major reason for the decrease in livestock mobility in Kazakstan. There is a direct relationship between the numbers of sheep owned by people and the distance they are prepared to travel. This is shown for former *Sovkohz* S. This *sovkhoz* had access to a small area of winter pasture on sands 100 km away from the farm property. Those new private farmers with 300 or more sheep tended to continue to make the journey, but those with fewer animals usually moved a shorter distance or did not move at all. The costs incurred in increased winter feed by making this migration are as follows:

Sovkhoz S had access to good winter pasture on a sandy massive to the north-west of Betpak-dala. In 1998 migration to this area had virtually stopped, resulting in increased expenditure on hay over winter.

Interviewees who did not migrate generally stated that they had to provide a minimum of two kilograms of hay per day per sheep for four months (120 days) in winter. Most people at *sovkhoz* S lacked machinery and so bought hay at a price of 1800 tenge per tonne. Grain, at a price of 5500 tenge per tonne was far too expensive for most. Therefore one sheep cost 432 tenge per year to feed. As an adult sheep was worth a maximum of 3000 tenge in Dzhezkazgan *oblast*, selling one enabled the farmer to buy enough hay to support seven sheep over winter. In contrast, a sheep kept on the sands in winter would need to be fed hay for up to six weeks, and would cost only 151 tenge per year. Despite these savings, migration to the sands is decreasing, and therefore it is clear that in this case the factors working against stock movement are more important than the advantages of migration.

A lack of animals is, however, not the only reason why mobility has decreased. Different types of farmers are developing grazing strategies that alter the decision regarding migration.

Emerging patterns: differences in grazing strategies among private farmers

There are two emerging types of independent livestock owner. The largest category is made up of those who did not receive or acquire enough animals or capital to start a viable private farm, or those who started one but lost their animals very quickly. These shepherds are basically subsistence farmers, selling a few animals from time to time (see Chapter 8).

The second group is made up of those who have higher stock numbers and access to a winter house, barns and wells, either as assets received on registration, or because these facilities are simply unused and available. This group has stable or increasing numbers of animals, and their farms could be described as commercial enterprises (see Chapter 8).

The difference between the two types of farmer is characterised by stock numbers. In Dzhezkazgan *oblast*, due to low meat prices and distances to markets, even flocks of 150 sheep were considered to be small and non-viable as the basis for a commercial enterprise. However, Kerven (Chapter 8) has found that in Almaty *oblast* the transition from subsistence to commercial farmer can be made at flock sizes of fifty, due to greater market access and better prices. Our case study looked at the behaviour of these two kinds of farmer in Dzhezkazgan *oblast*.

Behaviour of farmers with less than 150 sheep

Those people having very low numbers of animals are either private farmers who have become insolvent or families who never received enough animals to make it worth registering as private farmers. These families tend to graze their stock on common land around the village in summer, for which a tax must be paid. Such families also graze their horses near to the village, because in summer milking mares have to be milked every two hours. In winter the cows and sheep are kept in shelters most of the time, and the horses are sent out further from the village (normally about 15–20 km). Several farms have special common land set aside for horses.

The system for grazing household sheep usually involved about ten families taking turns to pasture all their sheep as one flock. For grazing horses a group of families share the cost of a shepherd, who takes the animals out every day or even in some cases lives out permanently in the steppe in winter, when the horses are pastured a long way from the farm. Few people pay established shepherds to pasture their sheep elsewhere, as half their lambs would be required to pay the shepherd.

Due to this system, the majority of stock remaining on the farms are now clustered around settlements. Those people with an intermediate number of animals also did not move to different pastures at all during the year. However they were normally based at a winter barn on the steppe all year round rather than in the village.

Behaviour of farmers with more than 150 sheep

In Dzhezkazgan *oblast* most people with larger flocks were registered as private farmers, and therefore had received specific areas of land for their animals, a similar situation as described in Chapter 5 for Almaty *oblast*. The amounts of pasture received by private farms in Dzhezkazgan *oblast* varied substantially – between 500 and 1800 hectares according to official records (Karaganda Zemliustroistvo 1998). However, these figures did not reflect the actual land use pattern. All respondents who grazed stock on the summer pasture said that there was enough land available not to be limited by anything other than distance from water. A practical reason for owning fewer hectares of pasture on paper was that tax was paid per hectare of land owned, rather than per hectare of land actually used. In practice people could go where they wanted, and lack of land was no obstacle to mobility, with access to wells, winter houses and barns (in a usable condition) being more of a problem.

Of those who had officially received pasture land, it was found that most were either not using this land at all, or had swapped it for other land so as to reduce their migration distances. It would seem that the benefits of the snow-free winter pastures or the productive summer pasture were outweighed by the costs of getting the animals there. Many farmers did not have access to transport, or even if they had vehicles they did not have the money for fuel or for the well pumps. However, even the richest farmers with access to all these necessities were often choosing not to go to distant pastures, and either swapped distant shares for those nearer the farm, or ignored the registration conditions and used abandoned pastures as near to their farm as possible.

There are therefore a number of factors, other than lower stock numbers, which are contributing to the cessation of migration. These factors are discussed below.

Access to services and markets

It must be remembered that the state farms in many ways facilitated stock mobility. Marketing of animals was conducted by the state and winter feed production was handled by specialised brigades. Children of shepherds could stay at boarding school free of charge whilst their parents migrated with the animals. Families would often spend the whole summer several hundred kilometres from the nearest town, but were supplied with goods and services by trucks from the *sovkhoz*. Since the recent reforms all these services were discontinued. In the study area shepherds preferred to be near farm centres in order to have access to links to town, both for marketing and for obtaining goods and services.

Stock theft

With the increasing isolation of those few shepherds migrating to distant pastures and the rising crime rate in Kazakstan, many people are afraid of moving far with their animals. There were reports of stock theft, and indeed this was the major reason why wealthier farmers have stopped moving to distant pastures in Almaty *oblast* (Kerven *et al.* 1998).

Alternative income from grain farming

In places where de-collectivisation was complete, *sovkhoz* hay brigades had been broken up and hay land distributed to private individuals. In such cases, those farmers with access to the equipment to cut hay themselves sometimes had difficulty travelling the distance between summer pastures and available hay land. On some farms the production of grain was possible, although it is marginal in all but the steppe regions of Kazakstan. Again, moving animals to summer pasture was sometimes sacrificed in order to spend more time on grain production either for direct sale or for winter fodder. For example, at *sovkhoz* D in northern Dzhezkazgan *oblast* there were 100 private farms registered, each with both ploughed land and pasture. It was found, however, that only seven of these were actually using their pasture land for grazing animals. Everyone else pastured all their animals on the common land around villages and either used their pasture share for cutting hay for these animals for the winter, or did not use it at all. Interviewees said that grain production was much more profitable than livestock production, despite the fact that zero yields can be expected every six or seven years. For these people, it was better to concentrate on grain production rather than to move out onto the steppe and live in yurts.

In both Soviet and pre-Soviet times, the costs and risks of migration were mitigated by the large numbers of animals being moved. In the pre-Soviet era Kazaks would have moved in groups of 20–30 families, pooling their livestock and other resources (Olcott 1995). In Soviet times, as in the case of the Moiynkum to Sary Arka migration, large numbers of animals would have been on the move together, and the maintenance of wells and other infrastructure was financed by the state. In post-Soviet Kazakstan, private farmers have found themselves very much on their own as most of the collective structures that have lingered since the reforms are now breaking up.

The kind of farms which eventually emerge as the norm in Kazakstan will determine future stock mobility. Migration, and in particular the very long distance movements, depend either on people with relatively large amounts of capital at their disposal, or on extended families with access to family labour, not for the migration itself but for the activities which support it.

Consequences of reduction in mobility on sheep nutrition

Livestock moved seasonally have access to different types of pastures with associated differences in nutritive value, as discussed in Chapter 3. This is evident from the changes in the energy content and yield of the various types of pastures over the year. The reduction in mobility therefore has implications for the quality and quantity of feed available for sheep. First, we consider the implications of no longer using the summer pastures of Sary Arka for livestock kept in the south of the study area. These animals now spend all year in the desert zone. Second, the forage conditions for those animals now kept in the semi-desert zone all year round are discussed.

Farms in the desert zone

The forage quality varies between ecological zones. The energy that is available to feeding animals from plants is highly dependent on plant digestibility. Table 7.1 summarises the vegetation types of the different seasonal pastures in the study area. Both this and plant fat content have been used by Ospanova (1996) to calculate the metabolisable energy per kg of dry plant matter of many plants found on Kazakstan's pastures over the growing season. Figure 7.5 shows these data for the most common species in the study region. There was not space to include a large number of species in this figure, however common species of *Artemesia* such as *A. pauciflora* and *A. sublessingiana* show patterns similar to that of *A. terrae-albae* (shown). *Festuca valesica* and *Stipa sareptana*, important components of the semi-desert and steppe zones, exhibit patterns similar to that of *Stipa capillata* (shown).

In spring and summer, those plants having the highest energy contents per unit weight are *Stipa capillata, S. sareptana, Festuca valesica*, and *Artemesia* species. There are advantages of moving sheep north from the desert to the semi-desert in spring to graze areas dominated by these species. Some of these plants become less edible in summer. *Stipa sareptana* is harmful to sheep after flowering in June, as is *Stipa capillata*, which can cause skin and wool loss (Kazgiprozem 1988). However the number of alternative high quality fodder species is high.

Table 7.1 Summary of different vegetation types and seasonal pastures in the study area

Season	Location	Vegetation type
Spring	Desert (clay) Betpak-dala	*Salsola arbusciliformis, Atriplex cana, Anabasis salsa, Ceratoides papposa, Artemesia terrae-albae, A. turanica.*
Summer	Semi-desert Sary Arka	*Artemesia terrae-albae, A. Lercheana, A. sublessingiana, Stipa capillata, S. sareptana, Agropyron fragile, Festuca valesica.*
Winter	Desert (sandy) Moiynkum desert	*Haloxylon aphyllum, H. Persicum, Calligonum spp.*

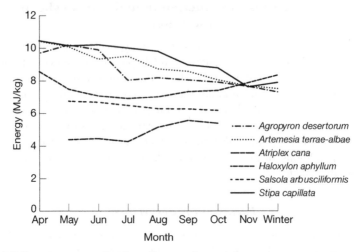

Figure 7.5 Energy contents of various plant species over the year.

Source: Ospanova 1996.

Plants typical of the drier and more salty areas in the southern desert have low energy values; these include *Atriplex cana*, *Salsola arbusciliformis* and *Haloxylon* species. The *Artemesia* species also found in this zone (*A. terrae-albae*, *A. turanica* and *A. pauciflora*) are less edible in summer due to a build-up of ether oils (Kirichenko 1980) and cows avoid *Artemesia terrae-albae* totally during this season (Ospanova 1996). Therefore at certain times of the year there is little alternative to the low quality saltworts. Given that these have, in some cases, 40 per cent less energy content than those species found further north which are edible in summer, sheep grazed in the desert zone would be expected to put on far less weight over summer than those further north.

The importance of the winter pastures in the Moiynkum desert can be seen by looking at the data for *Haloxylon aphyllum*. Towards winter, the absolute digestibility and thus available energy content of this species rises. The value of this and other *Haloxylon* species as winter fodder is therefore due not only to their shrubby morphology, but also because they have high digestibility at this time of year.

Previous animal migration patterns took advantage of seasonally-changing forage quantity as well as quality. Variable annual precipitation results in fluctuating biomass quantities and thus available forage for livestock. Figure 7.6 shows typical ungrazed biomass yields over the year for poor and average rainfall years in the desert and semi-desert zones.

In the desert zone the data are for vegetation dominated by *Salsola arbusciliformis*, the most widespread vegetation type. Statistical relationships between rainfall in winter and spring and peak biomass for this vegetation type were derived using biomass and rainfall data from Betpak-dala meteorological station (in Southern Dzhezkazgan *oblast*) and data collected in the same area by Kirichenko (1966). This

142

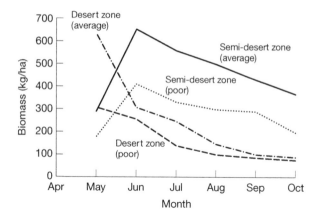

Figure 7.6 Biomass yield of different pasture types in the study area over the year.

Sources: Semi-desert zone: Kazgiprozem (1988), Koktas meteorological station.
Desert zone: Beloborodova (1964), Kirichenko (1966).

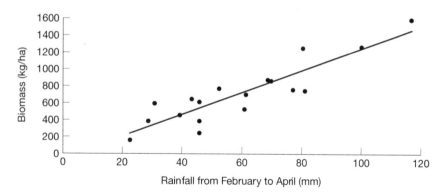

Figure 7.7 The relationship between spring rainfall and peak biomass for *Salsola arbusciliformis*-dominated communities in the desert zone. R^2=0.82, standard error of the estimate = 160 kg, N=18.

Sources: Kirichenko (1966), Betpak-dala meteorological station.

relationship is shown in Figure 7.7 and was used to estimate biomass production over 30 years from longer-term rainfall data. It was thus estimated that a poor year such as the one shown in Figure 7.6 can be expected to occur once every six to seven years.

In poor years biomass in the desert zone falls to 100 kg/ha or less by August, and even in average years it is not much higher. Such low yields would be unlikely to support large herds around single water points for the whole growing season without damage to the pasture and malnutrition of the animals.

It should be borne in mind that the data discussed above are for pasture which was not grazed, and which had a much higher biomass than grazed pastures which are dominated by *Artemesia*, *Atriplex* or *Ceratoides* species (Kirichenko 1966). Thus the data shown can be taken as absolute maximum biomass production in the desert zone for average and poor rainfall years.

In the semi-desert zone the biomass data shown in Figure 7.6. are for mixtures of *Artemesia* and *Stipa* species. Here, the lowest yields of less than 500 kg/ha occur only once every fourteen years according to the 30 years of biomass data from Koktas meteorological station (situated in northern Dzhezkazgan *oblast*). The frequency and severity of very dry conditions is much lower in the semi-desert zone compared to the desert, and available biomass is more stable. In the period from June to September the semi-desert is therefore a particularly important area for livestock rearing.

It cannot automatically be concluded from the above discussion that because sheep are no longer migrating to different seasonal pastures they are in poorer condition. It is shown in Chapter 6 for example that sheep still going to high-quality summer mountain pastures were in no better condition than those remaining on desert or semi-desert pastures. Forage availability may have a smaller impact on animal nutrition when stock numbers are low because they can be guided to those areas that have higher than average fodder quality and quantity. In the study area such pastures are along the river Chu. If the number of animals were to increase again, the animals would certainly benefit from the use of other pasture types in summer which are now hardly used.

Farms in the semi-desert zone

In the north of the study area, on sheep raising *sovkhozes* established in the 1960s in Dzhezkazgan and Karaganda *oblasts*, animals are not moved from the semi-desert pastures throughout the year. This means that although access to high-quality pasture in summer is good, reliance on winter fodder is extremely high as sheep cannot spend any time on the pasture in winter. For example, on *sovkhoz* Z in northern Dzhezkazgan *oblast* snow cover exists from between 100 to 140 days of the year at a depth of 20–35 cm (Kazgiprozem 1981). During all this time the sheep must be fed indoors. In the first years of privatisation such conditions led to mass mortality of animals on some farms. *Sovkhoz* A in Karaganda *oblast* became unable to produce grain in 1995 due to a lack of seed or working machinery. In the following winter it lost 5,000 sheep to starvation in the space of a few weeks. According to interviews in this region, the grain harvest fails due to low rainfall every six to seven years leaving stock on such farms totally dependant on hay through the long winter. The nutritional implications of this type of winter diet are discussed in Chapter 6.

Conclusions

The value of migration has historically been a way of maximising livestock production by allowing the most efficient use of the semi-arid rangelands of Kazakstan. In order for private farmers to be able to conduct long distance migrations today, they must have access to a winter and summer dwelling, a barn and a vehicle. They need to have enough animals to make migration both necessary and worth the cost. The wells, both in seasonal pastures and on migration routes, must be maintained, especially deep wells with pumps for which farmers would also need to supply fuel. Private farmers also need enough labour at their disposal to produce winter feed, market animals, and maintain all the assets needed to support the migration.

At present these conditions apply to very few individuals in the study area. The longer that remote pastures remain empty, the more expensive it will be to use them again in the future as much of the infrastructure has already fallen into disrepair. The great majority of Dzhezkazgan *oblast* (which is larger than the United Kingdom) remains devoid of stock, and this land cannot be used for anything other than livestock-raising. Vast areas of land are also now unused in other *oblasts* where sown agriculture is impossible. In rural areas where there are, at present, very few ways of making a living, this appears to be a waste of resources. However, whilst grazing resources near villages are available because stock numbers are low, there is no reason to migrate when convenience and safety dictate staying near the village. If overall stock numbers build up, the benefits of moving in terms of increased fodder availability would increase. Thus the costs of migration may again become worth bearing, at least for those with large flocks for whom the unit costs of migration are relatively low. However, it may be true that, in the present economic conditions, pastoral economies in very remote areas are simply non-viable. The location and character of stock-rearing in Kazakstan will become increasingly dependent on transport links and markets rather than the ecological factors which have defined it in the past. The decrease in the use of remote seasonal pastures may entail an increase in food supplements and inputs in order to replace the nutrients that these pastures provided in certain seasons.

8

PRIVATISATION OF LIVESTOCK MARKETING AND EMERGING SOCIO-ECONOMIC DIFFERENTIATION[1]

Carol Kerven

Introduction

Creation of market relations is the cornerstone of the transition from a command to a commercial economy. But these relations cannot be legislated into existence. After the state ceases to be the principal purchaser of goods, entrepreneurs must forge new links between producers and consumers. Exchange values based on supply and demand have to be established. Potential entrepreneurs have to acquire confidence that supply and demand exists for a commodity, with profitable margins, before they can take the risk of entering the market.

In Kazakstan, the exit of the state as the main purchaser of livestock goods was enacted very swiftly and fairly completely. In the immediate aftermath, barter became the dominant means of exchange in the absence of market channels. The state no longer supplied food to urban areas, while rural producers had ever more urgent need for income as state organisations ceased to function. This vacuum was filled within several years by a rising group of traders. Dissolution of state livestock farms had left some individuals with sufficient capital in the form of livestock or vehicles to allow them to start selling animals on a commercial basis. Some of these individuals also became traders. For the majority of less fortunate ex-state farm workers, selling animals became their only means of survival.

Thus arose a pattern of market relations described in this chapter. It is found that while markets are firmly rooted, families owning less than a critical flock size are not able to profit from the new markets. Commercial development is also

1 Some of the material on which this chapter is based was obtained during a study carried out for the Global Livestock Research Support Programme, 'Integrated Tools for Livestock Development and Rangeland Conservation in Central Asia', by the University of California, Davis.

inhibited the greater the distance between areas of livestock production and the main urban consumption centres. Domestic demand for meat is now the main impetus for marketing, while former demand for wool, other fibres, milk and pelts has not been entirely replaced by new markets. This changing demand structure gives price signals to producers who accordingly realign their production systems, which however takes some time. The loss of markets for some livestock products, combined with the collapse of state processing facilities, has meant that the domestic value added by processing is largely lost. There are now opportunities for filling these niches.

Historical overview

Livestock marketing is not a new phenomenon in Kazakstan. During the pre-Soviet period of Russian administration, Russians and other outsiders began trading and settling in the region from the 1850s. In the north, where contact with Russian settlers was greatest, some wealthier Kazaks turned from sheep to cattle-rearing to supply Russian demands for beef (Olcott 1981). In the south, there was less contact with new Russian markets but by the 1860s Russian administration adopted land policies to encourage Kazaks to become sedentary, combining farming with livestock-rearing (Olcott 1981). By the end of the nineteenth century, most Kazak pastoralists had become dependent on growing winter fodder and producing a surplus of livestock to exchange for the goods introduced by Russian traders in the new towns.

At various periods in Soviet history, families were permitted and sometimes encouraged to keep private animals, and to deliver their surplus production to the state (Channon and Channon 1990). The centrally planned economies of the Soviet Union were organised around state purchase orders, in which annual production quotas were delivered by state farms to central processing and distribution agencies. Livestock products in the form of meat, wool, milk, and pelts were redistributed by state agencies within each Republic for domestic consumption, or transferred between Soviet republics according to higher-order Union plans. Markets were thus purely 'captive' (Spoor 1997a).

The break-up of the Soviet Union combined with the removal of state purchase orders meant that outlets were no longer ensured. Each production unit, whether a small-scale family unit or a former state farm, had to seek its own markets in competition with other suppliers. In this economic environment, only the fittest have survived the first deep shocks of reform beginning in 1991.

Policy changes affecting livestock marketing

Since the early 1990s, the state has withdrawn from involvement in marketing. Over the same period, state collective farms have been largely disbanded. The initial result of market liberalisation and decollectivisation on livestock production was an imbalance between supply costs and output prices. Whereas formerly inputs and outlets were provided by the state, these had suddenly to be obtained through the open market. In many instances, markets for inputs and outputs of the livestock industry were either non-existent or poorly developed. Producers had in most cases no previous experience of dealing with a market economy, but had to adjust their entire production management systems to working within such an economy. Many failed to adjust. Most large ex-state farms have failed because of inherited accumulated debts, inefficient management and inappropriate production methods. At the same time, an unknown number of small-scale newly privatised producers have failed because they could not break free from the cycle of bartering their most transferable assets – namely their livestock – to obtain necessary inputs and the means of survival.

A major effect of the post-independence reforms has been that the non-state livestock sector consisting of privately-held animals has risen greatly in proportion to the state livestock sector (see Chapter 5). Most livestock marketing is now from private family holdings rather than from state farms. National statistics on livestock and products sold by each sector are not available, however.

In the study area, in 1998 only 40 per cent of smallstock were still owned by large institutional farms (cooperatives etc.), while 90 per cent of cattle and 81 per cent of horses were privately owned. In 1986, prior to decollectivisation, the number of privately owned livestock in the study area was so small that official statistics were not even collected on them. Nationally, a similar pattern prevails. In 1986 the state owned 87 per cent of sheep, a proportion reversed by 1998 when private and personal holdings accounted for 88 per cent of sheep.

Methods of study

Fieldwork was done in three periods, starting in January 1998. The bulk of material was obtained between July and September 1998, mostly in Dhambul *raion* of Almaty *oblast* in Kazakstan. This *raion* covers most of the study transect described in the Introduction to Part I. A further field visit was made in September and October 1999 to Moiynkum *raion* in Dhambul *oblast*, a different region in south-central Kazakstan which lies between 500 and 700 km west of Almaty. The purpose of this third field visit was to compare livestock marketing practices and prices in a more remote area within Almaty *oblast*. The latter region contains Almaty city, which has the highest livestock prices nationally.

Interviews were carried out with livestock producers and farm administrators situated within each major ecological zone of the study transect. The sample included seven cooperative farms whose staff, including directors and technical personnel of cooperative farms, were interviewed, as well as village administrators from village centres of cooperative farms. In addition, interviews were carried out at three livestock or fodder markets in Almaty and at the regional administrative centre. Several urban-based key informants were interviewed.

Respondents were selected to include most types of livestock management now found:

- production cooperatives (3)
- joint-stock cooperative, with some state funding (1)
- research institution farm (1)
- state-registered private farmers (8)
- unregistered private farmers (12)
- employed shepherds of cooperatives, joint-stock and research institutional farms (24)

A checklist was used for interviews with producers. Not all information in the checklist could be collected in each interview. Respondents were sometimes unwilling to divulge certain information, for example, on how many animals they sold or prices obtained. However, any lack of numerical data is counterbalanced by the insights that some respondents offered on their strategies for marketing livestock, how livestock markets are controlled, and reasons why prices varied by season. This information is reflected in the analysis of the numerical data.

For interviews with cooperative farm and village administrators, a more open-ended approach was used. General questions were asked on the production and marketing of livestock by the cooperative farm, as well as specific questions on the quantities, prices and seasons of sales. Some administrators were reluctant to reveal their farm marketing strategies, and information obtained is therefore not always complete.

Fieldwork involved camping at the grazing areas and small villages where livestock producers were situated in particular seasons. In this way, it was possible to observe some of the livestock management and marketing practices, in addition to interviewing people.

Commercialisation of livestock farming

Prior to privatisation in the mid-1990s, very few individuals were in a position to market their animals or animal products on any scale. Reforms introduced by Gorbachev in the latter part of the 1980s to encourage private family-based production resulted in the creation of 'model farmers', who were given loans and grants of land, livestock and capital equipment (Gray 1990). Their surplus production could be sold in local markets or bartered for other goods and services. Some of these model farmers have continued successfully in the independence period, while others have not.

In view of the limited commercial base of livestock production at the time of market liberalisation, the extent to which a commercially-oriented group of farmers arose in some three to four years is remarkable. Commercial orientation encompasses a range of attitudes and practices in which the objective of production is geared to meeting market demand. A commercial orientation among livestock-keepers is qualitatively different from simply selling a disposable surplus (Behnke 1983; Kerven 1992). A commercially oriented producer understands a particular market niche and attempts to produce for that market. Sales are planned, timed to coincide with optimal prices; particular types of animals are selected for the market, and raised in a manner which makes them more marketable.

A commercial objective does not prelude dependence on livestock also for meeting family subsistence requirements. But the subsistence-oriented producer, by contrast, sells on an ad hoc basis, often under economic duress. He or she hopes not to sell, as sales prevent herd/flock accumulation and reduce economic security, but continually must sell to maintain an income.

Both types of marketing – commercial and duress – are now clearly defined within the study transect area of western Almaty *oblast*, which is accessible to Almaty, the nation's largest and most profitable market for livestock. In Moiynkum *raion*, however, distance to large urban centres is greater, and livestock prices are lower. Small-scale producers in Moiynkum can only sell locally to small towns, as the distance and thus the cost of transporting livestock to the nearest big city is prohibitive. They can realise only about half the price per animal to that obtained by sellers in western Almaty *oblast*.

Whether a producer is commercially oriented or not largely depends on the scale of operation. In western Almaty *oblast*, farmers having more than about fifty sheep can begin to think about planning their sales and adjust their husbandry practices accordingly. Those with very few livestock find this impossible. Farmers in the more remote areas need considerably more sheep before becoming commercial (see Chapter 6). The largest operations, where several hundred and up to one thousand sheep (or equivalent values of larger animals – cattle, horses or camels) are owned, now exhibit many of the characteristics of commercial producers in market economies.

Market segmentation and specialisation

Separate private market channels exist for various products – live animals, meat, sheep wool, goat hair, camel hair, karakul pelts, merino-type pelts, etc. Entrepreneurs usually specialise in handling only one of these products. Market chains exist for some products, in which small-scale buyers obtain the products from rural producers and then sell on at wholesale prices to retailers in the towns. These entrepreneurs are termed in Kazak as '*alup satar*' which means 'buy–sell'. They are middlemen who connect producers with consumers, taking a mark-up along the way.

There have also emerged specialists who buy up young or thin livestock from producers, fatten these and resell later at a profit to consumer markets. These are farmer-entrepreneurs, and their presence indicates that a stratified livestock production system is developing.

These two activities – middlemen and farmer-entrepreneurs – have started since market liberalisation in the early 1990s. At the production end, a new group of richer farmers is investing steadily by increasing their stock numbers, buying equipment, renting good pasture land, and improving production methods. These are clear signals that foundations are laid for an integrated modern commercial livestock industry; profitable and segmented along specialised lines. This applies mainly to meat at present, however. Demand for meat remains high in Kazakstan where meat is valued in the culture, while incomes among some sectors of the population are increasing due to the oil boom. Export channels for meat which were developing several years ago have ceased, mainly due to veterinary health problems. The capacity exists for expansion of the meat sector, both to supply domestic and international markets.

The prognosis for other livestock products is not so clear, and the situation for each commodity is different. Table 8.1 shows average prices received by producers in 1998 and two years later in 2000 for various products. Prices for some commodities (e.g. camel and goat hair, sheep skins) vary by quality. Most of these products (except butter) are retailed by traders in urban markets, at higher prices.

Wool

From the beginning of the 1990s to 1999, the market for wool within Kazakstan was very weak, as most former state wool-processing and manufacturing industries had collapsed. Some wool is retained for the home, where it is used for stuffing mattresses. Most wool is exported to China, but producer prices were low, reflecting the low world price for wool during the middle of the 1990s (Kerven *et al.* 2000). In 1998, farmers did not consider income from wool as a significant contribution to their incomes, at a farm-gate price of 60 US cents a kilogram. As a result, most private farmers were switching from keeping wool breeds such as the dominant Merino-Kazak cross-breed in western Almaty *oblast*, to purely meat breeds, in particular, the hardy Kazak fat-tailed sheep.

However, by spring 2000, prices for wool collected in western Almaty *oblast* had risen 100 per cent from the previous year (Table 8.1). Over the same period,

Table 8.1 Producer prices for livestock products, Dhambul *raion*, 1998 and 2000

Product	1998 average price per unit sold (tenge)	2000 average price per unit sold (tenge)
Semi-fine raw merino, wool per kg	50	100
Rough Kazak sheep wool, per kg	No sale price	35
Camel hair per kg	50 (max. 160)	120 (max. 180)
Angora/cashmere type goat hair with down per kg	40 (max. 150)	180 (max. 500 for cashmere type)
Cow's milk per litre	12	
Mare's milk per litre	100	
Butter per kg	250 (max. 280)	
Sheep pelt (depends on quality)	100 (max. 600)	200 (max. 400)
Horse pelt	100	
Camel meat per kg	175	

the price of meat in Almaty had not risen, which meant in real terms the returns from selling meat had declined, due to inflation caused by the devaluation of the Russian rouble in late 1998. In response to this price differential between meat and wool, larger-scale livestock producers were planning to upgrade their flocks by introducing finer-wool Merino type sheep. Such individuals were aware of the higher price of wool in neighbouring Xinjiang province of China, and planned to sell there directly.

The improved wool prices applied mostly to fine and semi-fine wool. The desert region of Moiynkum *raion* produces coarse or rough-wooled Karakul sheep. There, in 1998 and in 2000, larger-scale livestock owners no longer even tried to sell their wool, and stored or threw away their wool crop. The cost of transporting this low-value coarse wool to market was not equal to the returns.

Pelts

There is a demand for sheep pelts to make the warm coats used by rural people in winter. A factory in Almaty makes these coats, but most pelts are exported to China, as prices are up to double across the border. Middlemen buy from Kazak producers and sell onto Chinese businessmen either in Almaty or at the border. Some Kazaks also cross into China to sell directly to Chinese factories.

Milk and dairy products

Strong local demand for milk and dairy products has led to some small-scale entrepreneurs, usually women, establishing regular deliveries of cows' milk from villages within a radius of Almaty, to the city. Due to the perishable nature of milk, and lack of refrigeration facilities, this marketing tends to be from villages not further than four hours' drive from Almaty. Income from sales of milk is particularly important to very poor families who may only have one or two milk cows and few other stock. Further from urban centres, within-village sales of milk and home-made butter take place, by women. Camel and horse milk is fermented at home for sale by farmers specialising in these animals, and is highly prized for medicinal properties.

With the collapse of state processing and distribution agencies, disposing of dairy products has become very difficult except at a local level. Urban marketing channels are poorly developed for these commodities, such that little reaches the urban markets. Formerly, state farms collected milk and fermented mares' milk for sale to institutional canteens during the peak production period over summer, but this service no longer functions. Interestingly, many producers, both large and small-scale, have ceased to milk horses and camels at all, and milk only enough cows for their family needs. This is because meat prices are now quite attractive, while it is difficult to market dairy products. From a producer's perspective, under these conditions it is economically advantageous to promote quicker animal growth through not milking. Consumer demand for dairy products is strong however.

Some European firms have already invested in dairies in Kazakstan, producing packaged milk products on a large scale for domestic sale.

Fibres

Animal fibres including goat and camel hair are sold by some producers. A cross-bred angora goat, with a fine, long fibre, is kept by many farmers, some of whom sell the goat fibre to itinerant entrepreneurs or directly at urban markets. Mainly, however, this warm fibre is retained for home use by knitting into garments for the cold winters. Camels are kept by only a few farmers, and the hair is also sold to middlemen, for resale either to a local small factory in Almaty or for export. The price received for camel hair is very low by world standards, and camel farmers do not receive significant income from this product.

The native Kazak goat produces a winter undercoat of very fine down. On the world markets, fine down of this type can be considered cashmere and has a very high value. However, during the Soviet period, the production of cashmere was not developed, and attention was devoted instead to increasing goat down yield by crossing the native goat with goats from other areas of the Soviet Union (Aryngaziev 1998). From 1998, the demand by China for cashmere rose, resulting in much-increased purchasing of goat fibre from Kazak producers. This is shown by the steep rise in prices between 1998 and 2000, up to a maximum of 3.50 dollars per kg from 0.50 dollars per kg in 1998 (Table 8.1).

Meat

By far the greatest proportion of income for all producers regardless of flock/herd size, whether individual or cooperative farms, comes from the sale of live animals for meat. Animals are sold either alive for immediate slaughter at the urban markets, or slaughtered and the meat transported to market. Other producers, either to fatten or to add to their inventory, buy a proportion of animals marketed. It is more profitable to sell an animal alive for slaughter, as buyers prefer to see the animal and particularly value animals with more fat. However, poorer farmers cannot afford the transport costs of taking animals to market, and therefore slaughter at home and take only the meat to be sold at market.

Variation in livestock prices

Sales of animals for slaughter take place through the year, but more commercially-oriented producers time their sales to periods when market prices are highest. Poorer producers sell a few animals frequently, sometimes once a month, while larger commercially-oriented farmers only sell once or perhaps twice a year, in large batches.

Figure 8.1 shows sheep prices over six months in Almaty markets. The best prices occur in early spring (February–March), but only for fattened animals. The

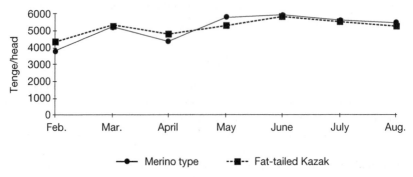

Figure 8.1 Almaty market sheep prices, 17 February–5 August 1998.

Source: TACIS Agro-Inform 1998.

prices for fat sheep in January 1998 averaged at 10,000 tenge, dropping to 6,000 tenge by July in the same year, observed at Almaty markets. Fewer animals are offered for sale in winter, as the bulk of smaller producers try to retain their animals over winter until they can be fattened on free, natural pastures in the spring. Those animals sold by small-scale producers in winter are thin and underweight, and therefore not attractive to consumers. In March there occurs an important national holiday period of the Kazak New Year (*Naarus*) when families try to slaughter animals for feasting. If producers can sell a fat animal at this lean period, they can obtain a premium price. This is exactly the strategy now adopted by the more commercially astute farmers. While most producers are aware of the price differentials by season, only larger-scale farmers can take advantage of these differentials by capturing the scarcity value of marketing fat animals when few are for sale.

Larger-scale farmers can obtain the better prices by being able to afford better quality and quantity of feed over winter. They select only young male animals (six months of age for sheep) which are kept in barns for up to six months before and during winter, fed on grain and the more nutritious cut grasses. Good quality fodder is now relatively expensive, and poorer farmers have to reduce costs by feeding their stock on whatever is cheapest, and letting the sheep graze in the open as much as possible over winter (see also Chapter 6). Farmers are generally aware of the nutritional benefits of different fodder sources. However, poorer farmers cannot use this knowledge to raise the output of their livestock and thus realise a greater return on their livestock through tactical marketing.

The period when livestock prices are lowest is late summer and autumn (August–October). The reasons are as follows. Most animals offered for sale have gained maximum fatness by grazing over the spring and summer, and therefore a 'buyers' market' prevails, in which buyers have a lot of choice. Many small-scale farmers try to sell their surplus animals at this time, as they need cash for winter food supplies and necessities for children starting the school year. These farmers also wish to avoid keeping surplus animals alive over winter as this requires feeding supplementary fodder. Flocks and herds are culled, and the market is glutted. Commercially oriented farmers, on the contrary, do not sell at this period but are instead buying cheap fodder from the new harvest and preparing to fatten up their young animals over winter.

Overall, prices for livestock sold in Almaty increased from the mid-1990s up to 1998, but then levelled off in 1999, due in part to an influx of animals brought in from Russia in late 1998 and early 1999. This was the period of particular economic hardship in Russia as the rouble collapsed. Markets in Kazakstan were flooded with animals being sold at depressed prices.

Prices obtained for animals decrease with distance from Almaty market. This is shown in Table 8.2 in 1998. Similar patterns prevailed in 1999. The cost of transporting animals by vehicle over long distances is not economic for smaller-scale producers. Traders recognise that producers in remoter areas are willing to sell at lower prices, and will visit these areas to buy up cheap animals for resale in the cities.

Livestock and meat prices are higher in Almaty than for most other areas of the country. Prices in several other cities in summer 1998 are shown in Table 8.3. While average sheep prices per head are not always highest in Almaty, this does not necessarily reflect the price by weight. The weight of different sheep breeds varies, while different breeds dominate in each region. An indication of the price per weight is given by the meat prices. Table 8.3 shows Almaty having the highest meat prices, though Astana, the new capital had even higher meat prices – 275 tenge/kg for mutton – at the same period, but not recorded here (TACIS 1998). The market prices for beef versus mutton are affected by demand. Beef prices are higher in cities with larger proportions of Russian or foreign residents (Almaty and around the Caspian oil and gas industry in western Kazakstan).

Flows of livestock to different markets vary by demand and proximity to particular cities. Thus southern Kazakstan sheep from Kyzl Orda and Shimkent *oblasts* are mainly sold in Tashkent, Uzbekistan. Sheep and cattle from western (Aktyubinsk and Uralsk *oblasts*) and northern Kazakstan (Kustanai and Pavlodar *oblasts*) are mainly sold to Russian centres across the border. Almaty city is supplied by the following *oblasts* in addition to Almaty *oblast*: Karaganda, Dzhezkazgan, Semipalatinsk and Ust-Kamenogorsk.

Table 8.2 Prices for adult male sheep in 1998, by distance from Almaty

Location of market	Price in tenge (79 tenge= 1 US$)
Small village in Moiynkum *raion*, 600 km from Almaty	2,500
Small village market 350 km from Almaty	3,000
Uzunagach market, 50 km from Almaty	5,000
Almaty market	8,000

Table 8.3 Average prices for livestock and meat in some cities, August 1998

Average market price in tenge per head	Aktyubinsk (western region)	Almaty	Taras (Dhambul, southern region)	Ust-Kamenogorsk (eastern region)	Shimkent (southern region)
Sheep, fine-wool	nil	5,500	3,900	5,500	5,000
Sheep, fat-tail	4,500	5,200	5,600	nil	5,700
Sheep, semi-fine wool	3,100	5,000	2,600	5,500	4,700
Sheep, karakul	4,800	nil	3,900	nil	4,000
Bulls, meat type	42,500	29,000	27,700	32,500	28,300
Cows, meat-milk type	24,000	27,000	28,000	27,500	24,800
Horses, meat type	28,500	39,300	28,000	27,500	29,700
Mutton, tenge/kg	181	270	225	240	197
Beef, tenge/kg	200	275	201	245	175

Source: TACIS Agro-Inform 1998 (all towns for which data available are given here).

Marketing and subsistence strategies by types of producers

Shepherds are different – one has, another does not have.

(Older shepherd, Dhambul *raion*)

Livestock-owners now belong to several distinct groups, each with a particular marketing and subsistence strategy. One of the key differences between producers is in the number of animals owned. The greater the number of animals, the more a producer is able to plan sales on a commercial basis.

Several social and economic factors influence whether an owner is likely to have more animals. These include the owner's past and present employment status, whether they have inherited animals, and the age of the family head. More successful livestock-owners tend to be those who formerly had or still have a good position as a senior shepherd or professional in the state farm (*sovkhoz*). These people often received disproportionately large shares of animals and assets upon the break-up of their *sovkhoz*, as a reward for their past services. If one of the fortunate few still employed by the newly created agricultural cooperatives, an individual has preferential access to key cooperative resources such as winter fodder, winter shelters and transport.

By contrast, many of those who are unemployed received little or nothing from their former *sovkhoz* upon its dissolution. These people have no additional sources of income, for example, from renting out agricultural equipment or trucks. They have great difficulty retaining a sufficient number of private animals to become commercialised. There are exceptions to this pattern, of people who are now prospering through their own initiative but were considered 'lazy' workers for the *sovkhoz*, or were not socially well-connected to the *sovkhoz* management and therefore did not receive a share of the assets upon the *sovkhoz* dissolution.

The Kazak practice of pre-inheritance, whereby adult sons and married daughters may be given a portion of the senior generation's private animals, assists in building up an individual's flocks and herds. Older livestock-owners (past thirty) have had a longer period to accumulate animals over their lifetime. Therefore, more commercialised livestock-owners tend to be more than 40 years old. Exceptions occur, in which a young man has inherited or been given a significant number of animals by his father or father-in-law. Thus luck plays some role in determining whether a livestock-owner is likely to become commercially-oriented.

Ultimately, the more commercialised livestock-owners are successful due to a combination of several factors, but key ingredients are adaptability to new economic circumstances and a will to succeed. Shepherds can tell many tales of families who have lost most of their animals after decollectivisation, due to bad management, drunkenness, laziness or sometimes bad luck. In the harsh opinion of one senior technical worker in a *sovkhoz*:

> As for poor people in the new system, one is poor who does not work. In the old system, the state was controlled by the rich – even now. Poor people are lazy, stupid and drunkards. In the future, one rich *bai* [nobleman] will eat up all drunkards and make them his slaves. Nature will make regulation and establish order again.

Animal products can contribute to families in a number of ways: directly as food, clothing or transport, and indirectly when exchanged. Animals or their products can be exchanged through a market in return for cash or goods, but they can also be exchanged through a social network, giving value back to the owner by building up reciprocal links between individuals. The ways in which animal products are used or exchanged varies markedly between different types of producers.

The livestock holdings and sales of three groups of livestock farmers are shown in Table 8.4. Only a sub-sample on whom sufficient quantitative data could be obtained could be included in this table. Further details about each group's characteristics were obtained, as discussed below.

Group 1: private large-scale (n = 6)

Private large-scale livestock owners are self-employed, with more than 100 head of smallstock. Three of the six owners in the sample are older men in their sixties, ex-head shepherds of their former *sovkhoz*. Of the other three, one man ran a *sovkhoz* shop in the past, another is 25 years old and just inherited several hundred animals, while the third is in his thirties and currently employed as a guard by a government institute.

In the sample, four out of six in this group had registered private holdings of pasture land or former *sovkhoz* barns. The other two are guarding barns belonging to institutes in grazing zones distant from villages. Four out of six of these farmers

Table 8.4 Livestock ownership and sales by producer type in study area (n = 25)

Ownership in July 1998 and sales July 1997–8	large-scale private (n = 6)	small-scale private (n = 10)	employed shepherds (n = 9)
Mean no. smallstock owned	390	27	61
Range	110–1,000	5–70	10–200
Mean no. cattle owned	14	3.5	12
Range	10–22	2–10	1–18
Mean no. horses owned	11	1.5	9
Range	3–17	0–4	2–18
Camels owned (1 farmer)	23	n/a	n/a
Mean no. smallstock sold in last 12 months	100 (n = 3)	6.4 (24% of flock)	24 (39% of flock)
Mean no. cattle sold in last 12 months	n/a	0.8	1.1
Mean no. horses sold in last 12 months	n/a	0.1	0.9

are managing livestock in multi-generational family groups, typically consisting of an older father, retired from employment as a *sovkhoz* shepherd, together with his adult sons or daughters.

The women in these families are jointly responsible for processing milk and wool from the group's livestock, while the men (including sons-in-law married to daughters) take care of herding and marketing the animals. The other two men in this sample are partners in a family group of male cousins, who have combined their livestock and other resources. Two of the six have also hired assistant shepherds, not related. The family members in these livestock-owning groups are not necessarily co-resident. Married children have usually acquired their own houses, though often in the same village as their parents. However, male cousins in a livestock management group may be living in separate villages, while co-managing livestock.

Marketing by private large-scale farmers is planned on a commercial basis. Young male animals are selected for fattening over winter, fed on the best quality of available feed (grain and more nutritious cut hay), and sold in spring time. Old and barren females are also sold. These farmers compare livestock prices in the two largest markets in the study area before taking animals to sell, and choose the market with the highest prices. They have their own trucks for transporting animals to market, and may also own land for growing fodder crops, as well as owning agricultural machinery. Most of these farmers in the study sample were also buying up young stock at markets, in order to fatten them for resale later.

There are distinct patterns in the way large-scale farmers dispose of other animal products. Since their larger herds of cattle produce more milk than is needed by the family, only a few cows are milked and calves are allowed to have all the milk from the rest. This will result in a faster rate of maturation among the calves, and is a commercial approach. Such families never sell milk products, but surpluses of cream and cheese are given away to relatives. Likewise, milch-cows may be loaned to poorer relatives. Such gifts develop bonds of reciprocity. It is expected that richer relatives will assist poorer ones, in return for which they can expect contributions of labour (Werner 1998).

Group 2: private small-scale (n = 10)

These livestock owners are typically unemployed or retired, for whom livestock provide some of their subsistence. They struggle to maintain animals if they have no other source of income. Animals are mainly used for subsistence; milk from cows, home consumption of smallstock, and home use of wool and pelts. Owning small numbers of stock, these owners cannot develop the commercial approach of the larger-scale farmers. Within the sample, four were unemployed, four retired and two were still employed guarding barns for cooperatives. Pensioners are the best-off among the poorer farmers, as they still have a cash income from the state pension. The pension often comes late, which is the main reason why pensioners sell an animal. Pensioners usually have assistance in managing animals from some of their adult children, who may pool their own animals with the parents' livestock. The unemployed can use their horses or donkeys to do odd jobs pulling carts. For the unemployed with very few animals, dependence on remittances from other family members is necessary. This group depend much on selling or using milk from their cows, especially among those having very few smallstock left to sell. Only half the sample was in multi-family livestock management groups, in contrast to richer private farmers, all of whom manage their livestock in larger family groups.

Small flocks cannot provide a sole source of income for such families. They can farm animals out to relatives working as shepherds for cooperatives, thereby avoiding the labour cost of tending a small flock and also have their animals fed with fodder supplied by the cooperative. These families also use neighbourhood groups for livestock herding. Their main difficulties are obtaining enough winter fodder, obtaining transport to market and, if distant from Almaty, receiving lower prices at local markets.

Group 3: employed shepherds (n = 9)

The sample included seven sheep managers, one horse and one camel manager. The age of the household head ranged from mid-twenties to mid-sixties. Faced with the need to lay off most of their shepherds after the *sovkhozes* were dissolved, the new cooperatives kept on only the very best livestock managers. Within the small villages that used to employ all able-bodied residents as *sovkhoz* workers, these remaining employed shepherds are now a privileged few. Food security is the main benefit from employment as a shepherd, as one young man who recently lost his job commented: 'I would prefer to be a shepherd as you have everything in your hand; milk, butter, meat. You can eat at any time.'

Employed shepherds are better-off than private small-scale farmers as they receive winter fodder at cost, and sometimes receive veterinary treatment and transport, all provided by their cooperatives. They have rights to graze their own animals on grain and hay fields belonging to the cooperative, and to harvest crop residues and hay from this land for their own stock over winter. As they receive food and other benefits from employment, they are not always forced to sell live animals at a disadvantage. Their livestock provide subsistence, as the family does not have to buy animal products. Animals and products such as cream, wool and pelts are sold to purchase necessities such as tea, sugar and clothing, as shepherds receive no cash wages from cooperatives. These shepherds can also help out other family members by taking their animals to the better pastures belonging to the cooperatives. In return, the shepherd family can expect these relatives to assist with livestock management.

Some employed shepherds are already making the transition to becoming private large-scale livestock farmers. These shepherds are buying up lambs, foals or young camels at below-market prices from cooperatives which are failing financially. Some cooperatives pay their shepherds in young animals, geared to the level of offspring which a shepherd manages to raise from cooperative animals in a season. The immature animals are grown on by the shepherds and re-sold on the market at a profit. Such shepherds have also joined in larger family livestock groups, buying a 'share' of a barn, truck or agricultural machine. Many employed shepherds do not raise animals on a commercial basis, but sold them to acquire the better things in life. As an older shepherd remarked: 'I am selling my animals now; that's why I'm still alive, and educating my daughters and have nice furniture.' Their flocks are large enough that animals can be regularly sold without depleting the breeding stock. However, some are hoping to establish themselves as private commercial farmers in the future, when their family livestock business is up and running.

Emerging market chains and restrictions

Private entrepreneurs have now replaced the former centralised state marketing mechanisms. Livestock traders buy from and sell to particular areas, seeking the lowest purchase prices and highest retail prices. Sellers at the Almaty markets are either fairly large commercially oriented livestock farmers or professional traders. Some traders come from as far away as Semipalatinsk or Dzhezkazgan *oblasts* in the north, bringing up to 100 sheep by truck every week to sell. Almaty livestock markets also draw traders and private farmers from Dhambul and Karaganda oblasts, up to 1000 km distant.

As prices for livestock and meat at Almaty markets are among the highest in the country (see Table 8.2), traders are attracted from distant production areas to sell in these markets. The marketing situation for producers, traders and consumers in Almaty *oblast* is atypical for the country as a whole. Almaty city undoubtedly contains more better-off families than other areas of Kazakstan, due to the economic boom created by the oil and gas industries, having more foreign organisations and better employment opportunities. Almaty's higher-income families can afford to pay more for meat, while by contrast the loss of employment in other cities results in lower demand and prices for meat.

Within the study area, live animals are bought by small-scale buyers, who form family groups composed of a father and son, several brothers or close male kin. Women are not involved directly in trading live animals, but are mainly involved in selling wools and milk products. Livestock traders are typically from a rural background, involved with livestock. As one young man selling sheep in an Almaty market said: 'We used to ride horses, now we sign contracts'. Trading in animals is more financially rewarding than trying to subsist on a small private flock. However, as discussed below, access to trading is restricted.

As transport is the crucial element to marketing, to become a buyer requires owning a truck large enough to transport animals and fodder. The new trading family groups often include ex-*sovkhoz* drivers who received a truck as part of their severance allocation when their *sovkhoz* was dissolved.

Livestock traders attend the weekly rural markets and buy between ten and thirty sheep, purchased individually from private owners. Sellers stand next to their animals at these markets, while prospective buyers walk around inspecting the animals on offer, trying to strike a price. Some traders also buy directly from producers, travelling around in trucks to villages. The traders have several strategies for selling on animals. One kind of trader resells their collected animals directly by truck to larger, more profitable markets, the main ones being in Almaty. Other traders retain thin and young stock they have bought from sellers disposing of animals due to economic necessity and fatten these up for a month or two before reselling at the larger markets. Buyers particularly choose to buy animals for fattening in late winter for resale in the spring, as animals sold in winter by poorer families have usually been nutritionally stressed, and will be sold cheaply.

Traders can hold up to 200 sheep for fattening. More than this is considered too financially risky, as resale prices may not cover feeding costs. Also few people have barns that will accommodate more than 200 sheep for fattening. The barns have been privatised by gift, rental or sale from the former *sovkhozes*. The animals are stall-fed on different feeds, including *Agropyron*, lucerne, mixed mountain hay, barley grain, maize silage or wheat meal. *Agropyron* is the most commonly used feed for this purpose in Almaty *oblast*. Since privatisation of state farmland in the mid-1990s, there has been a rise of private farms growing fodder crops for sale in the vicinity of Almaty, to supply the commercial livestock sector. Most barns used by traders for fattening animals lie within the foothills or semi-steppe southern part of the study area, where higher rainfall permits cultivation (see also Chapter 5).

Traders sometimes work in partnership with friends or relatives who have land on which fodder crops are grown. An integrated enterprise of trading animals and supplying fodder increases profitability. Success in trading, as in managing livestock, is partially the result of links between family enterprises.

There are some institutional restrictions on being able to sell animals or meat. The principal livestock markets in Almaty *oblast* are in Almaty city. There are three: Razvilka and Kamenka, which are retail markets, and a wholesale market. Live animals and meat are sold every day at these open-air markets. The markets are on main roads at the outskirts of the city. Access for sellers to these markets is controlled by traffic police, who check vehicles bringing animals in from the countryside and charge fees (both official and unofficial) to those seeking to sell at the markets. Information on the rates for fees could not be obtained. City officials control the markets by charging fees for using animal pens or stalls to sell meat. A fee of 150 tenge (about $2) per sheep sold was mentioned by sellers.

The Almaty markets are thus not open to all sellers, in contrast to the regional centre market at Uzunagach (Dhambul *raion*). Only those traders who have established links with the city officials are permitted by them to sell from Almaty markets. This is a new system, which is referred to by rural shepherds as 'the racket'; groups of traders in collaboration with city officials control prices and access at the city markets. A rural seller can avoid paying some of these charges if he has family connections to an important person in the city. This protection is known as 'an umbrella' (and also includes selling of other goods from kiosks and market stalls). If not protected through social connections, a trader has to pay into a 'racket' to be allowed to sell in the markets. Rackets are said to be run by very well-connected and wealthy people, some of whom also control the roads on which charges are levied for transporting livestock to Almaty markets.

Livestock producers sometimes avoid selling their animals in Almaty because of these restrictions. Although prices are usually higher in Almaty, producers selling only a few animals at a time find that the 'fees', which have to be paid in order to sell animals in Almaty, do not justify the higher city prices. Sellers at Uzunagach market have to pay about 100 tenge ($1.25) per sheep sold, and access is not restricted. Anyone may bring his or her animals to sell at the market place.

Constraints to marketing

The main constraints to marketing are experienced by small-scale family farmers who live at some distance from the main urban markets. Whereas in the past all animal products were distributed through state organisations, each producer must now find a market. Good prices can be obtained in the towns and cities for animals destined for domestic consumption. However, producers who can only sell a few animals at one time and live several hundred kilometres from an urban market find the costs of transport to main markets prohibitive. Instead, they sell at small local markets. As demand for meat in rural areas can usually easily be met, rural market prices are low.

By contrast, large-scale private producers and livestock cooperatives can benefit from their size by bulk marketing. These producers can use high-capacity trucks carrying seventy or more sheep at a time. Large-scale producers have few constraints to selling live animals at this time, apart from the restrictions on access to certain markets.

Local community transport groups have not arisen for marketing animals of small-scale producers. Individuals who own transport do rent out their vehicles to neighbours wishing to sell animals in the urban markets. But producers have not formed joint livestock marketing associations beyond the immediate family. The reason often given is that too much trust is required to sell one's animals through a larger group. However, some villages have developed small marketing groups for selling milk, run by village women with kin links to milk-sellers in the cities. Though not without problems, livestock marketing groups have been successful in other pastoral areas of the world.

A second constraint to better marketing opportunities has been the collapse of industrial processing facilities in the immediate post-independence period. At the end of the Soviet era, Kazakstan had three large wool washing facilities, four wool spinning mills and two wool fabric mills.[2] These factories processed 77,600 tonnes of wool from Kazakstan in 1989, as well as wool from neighbouring Soviet Central Asian countries. In the first half decade of transition, these factories either closed or were reduced to a fraction of their former throughput, as both wool production and export demand fell drastically. Production of washed wool by 1995 was one tenth of the 1989 level (ADB 1997). By the end of the decade, some of these factories had received foreign investment and advice as well as new technology. These factories had begun processing wool again, albeit a much lower volume compared to the Soviet period. By 1999, about half of Kazakstan's wool production of 30,000 tonnes was being processed domestically, with the remainder exported raw, mostly to Russia. Domestic processing of skins and hides also virtually ceased in the 1990s, due to financial difficulties within the state-owned leather factories. Thus, marketing of hides and skins was constrained.

The collapse of domestic processing facilities for livestock products had a direct and negative effect on the demand for raw products – wool, fibre, skins and hides. By the mid-1990s, the vacuum was being filled by international buyers from China, Iran, Turkey and other countries. These new buyers moved into the domestic Kazakstan markets to purchase raw materials from producers desperate for income and therefore willing to sell at very depressed prices (ABD 1997). During the mid-1990s, the world price for wool fell sharply, though by the late 1990s prices for wools had risen again. By the beginning of the current decade, demand for finer wools is now improved, resulting in considerable price increases for producers.

Associated with this is the breakage of international market links that would give higher prices for Kazakstan's specialised animal products. While meat, milk and

---

2   The author is grateful to Nurlan Malmakov for the information about wool processing.

wool marketing is now quite well-established around centres of demand such as Almaty and other main cities, similar marketing opportunities for specialised fibres – fine Merino wool, goat and camel hair – did not arise until the close of the 1990s. The prices received by producers for these products were negligible during most of the 1990s, but rose with demand from China at the end of the decade, as Table 8.1 shows. However, commercial links for processing and retailing wools, pelts and fibres remain underdeveloped, in contrast for example to Mongolia and China, where a thriving privatised industry in camel hair and cashmere now exists.

Conclusions

Marketing is now firmly in the hands of private entrepreneurs. The state plays very little part in marketing livestock and their products, apart from taxing animal movement to market, inspection of animal products for sale and tariffs on wool exports. Police and veterinary personnel exert control, often involving bribery. As state interventions in marketing have rarely proved beneficial in other pastoral areas of the world (Kerven 1992), increasing the role of the state in marketing Kazakstan's livestock is not recommended. The state could, however, provide material support in identifying more profitable international markets.

Separate market channels exist for various products. There have also emerged farmer-entrepreneurs who buy up livestock from producers, fatten these and resell later at a profit to consumer markets, indicating that a stratified livestock production system is developing. These are clear signals that foundations are laid for an integrated commercial livestock industry; profitable and segmented along specialised lines.

The capacity exists for expansion of the meat sector, both to supply domestic and international markets. Niche markets for specialised livestock products (fibres and pelts) should also be developed. International technical advice is required to help identify profitable markets and to improve the quality of livestock product outputs to meet international standards.

A stratification is emerging among Kazak pastoralists. A new group of richer farmers is investing steadily by increasing their stock numbers, buying equipment, occupying good pasture land, improving production methods and marketing on a commercial basis. The majority of farmers remain subsistence-oriented, as they own insufficient animals to accumulate wealth through marketing or to invest into livestock production. The needs of these two groups, and thus the solutions to their constraints, are different.

Support targeted to small-scale livestock farmers and cooperatives is recommended

• Seasonal agricultural credit to small family groups, for spring cultivation of winter fodder crops. Credit is needed to buy inputs in spring: mainly spare parts, fuel, and seeds. The credit would be repayable in winter or the following spring, from the sale of yearlings.

- Transportation links could be developed for joint marketing by family or village groups. The cost of transport is a major constraint to smaller-scale farmers in marketing livestock to the major urban centres.
- Local micro-processing of livestock products, including pelts, dairying, meat processing, e.g. sausages. Techniques for improvement of home processing could also be introduced, for quality pelts and wool sorting.
- Reviving traditional handicrafts for weaving wool into carpets would increase rural family income from livestock. Livestock-owners in Turkmenistan, for example, are able to derive income from making and selling carpets, a skill which is gradually being lost in rural Kazakstan.

Support to all livestock marketing

- Development of specialised wool and pelt production and local processing. Potentially high-value wools and pelts already exist but currently obtain very low prices – including camel and angora wool and karakul pelts. Both large- and small-scale producers could be assisted to gain added value from these products, through improvements of breeds, livestock management and processing. Other high-value livestock, such as cashmere goats, could be economic and suited to the semi-arid environment.
- Development of export markets for specialised high-value products is necessary, requiring market surveys and feasibility studies.

Profitable marketing of livestock and their products is crucial to the development of Central Asia's rangelands and the people who still depend on the rangelands for their livelihoods. A sustainable offtake rate will relieve pressure on the natural resources, enhance the value of the rangelands to the nation and provide much-needed employment and investment capital for rehabilitating the livestock sector. As another observer on the agrarian transition in Central Asia has concluded: 'Much will depend on the formation of an accessible and competitive market environment in which these new (often family-run) farm units can operate' (Spoor 1997b: 2). In spite of neglect by the new national government, marketing systems have spontaneously developed to meet demand. Kazak farmers and business people have shown that they are capable of responding to commercial incentives, without any assistance. However, judicious external support is still required if livestock are to become the commercial asset that they have the potential to be.

Part II

PASTORALISTS AND RANGELANDS OF THE KARA KUM DESERT, TURKMENISTAN

The pastoralists in the study area reside in small isolated communities of between two and forty families. Each community is composed of one or more lineages, descended from Turkmen nomadic families that inhabited the desert before the Soviet period. While modern ethnographic accounts of Turkmen pastoralists are lacking, their economic and social systems in the pre-Soviet period are described in papers of the USSR Institute of Ethnography (1973). The desert communities are at least 10 km apart and often much further, connected by tracks through sand dunes. The basis of every settlement is a well tapping into groundwater, annually recharged with the spring rains. Rainfall is sparse, with an annual average of 150 mm and crop agriculture is not possible in the desert. Virtually all desert families raise livestock, mainly sheep but also goats, camels and cattle.

Desert households are almost always multi-generational as when sons marry they either remain in the same household as their parents or move into an adjacent dwelling. Livestock holdings of the father may be shared out between the sons or retained under the father's control, depending on individual families. Young women move into their parent's-in-law houses upon marriage, and work for their husband's family.

As in Kazakstan, there remain strong bonds of mutual assistance between kin, though rural–urban links between desert- and town-based relatives do not seem as pronounced as in Kazakstan, as fewer desert children go on to study and work in the towns.

The tasks of livestock management are divided according to age and gender. Young boys, and sometimes young girls, take animals on foot to be grazed around the villages. Older boys and men use motorbikes to herd animals further afield. Women remain close to the house and animal shelters, and take care of giving fodder to animals, as well as milking and watering them. Women and girls process milk into yoghurt and wool into yarn for clothing. Men take animals to market and purchase household food requirements in the cities.

Most families in these desert communities were employed as shepherds for the state farms (*sovkhoz*), centred in the south of the study area along the Kara Kum canal. The state farms constructed housing and schools for the desert communities, as well as lining wells and providing motor pumps for the wells. Electricity is provided to some larger desert settlements nearer the canal zone or the main road. There is no telephone communication. Remoter settlements do not have an electricity supply. Drinking water for humans is usually brought in by truck from the canal zone, and is now a major expense for families. The state farms provided winter feed for livestock, transport to distant pastures for people and water for animals, and fuel for water pumps and other mechanical equipment.

Residential arrangements centred on the small villages, from which shepherd men would move out to stay with the animals at remote wells in each season. The state farms did not provide their shepherds with housing at the seasonal pastures. Whole families moved out from the villages in spring time, staying at the distant pastures in tents or, more rarely, yurts. These patterns generally continue today, but with some changes.

The pastures of Turkmenistan cover about 90 per cent of the country, with an area of forty million ha. The pastures are mainly sandy desert, with annual precipitation of less than 250 mm. At the end of the Soviet period, one quarter of the agricultural output was derived from livestock, principally the Karakul sheep. Livestock made up a smaller proportion of the agricultural economy compared to Kazakstan.

Study areas

The study was conducted in two different areas, as a comparison. One area is occupied by the Yerbent peasant farmers' association, or *dihan birlishik*. The association's territory lies in the central Kara Kum desert, and begins 80 kms north of the capital Ashgabat on the main Ashgabat–Dashauz highway. The area stretches more than 200 km to the north, covering a band 50–60 km wide. Yerbent has close relations with Turkmen Mallory, the Government agency responsible for livestock associations. The administrative centre of Yerbent is at Bokurdak about 100 km north of Ashgabat. The farm was formerly a state farm until, in line with the reforms of the agriculture sector, it became an association.

The second site is adjacent and to the west of Yerbent association and the main Ashgabat–Dashauz highway, and is in Gok Tepe district (*etrap*). This district stretches from the border with Iran in the Kopet Dag mountains in the south, some 200 km northwards, stopping short of the boundary between Ahal and Dashauz regions (*velayat*).

In the south of the study area lies the canal zone, where settlement is relatively dense, in villages which are also former *kolkhoz* or *sovkhoz* (collective farm) centres. Crops grown under irrigation are wheat, grapes, melons, vegetables, cotton, barley and lucerne as fodder, mostly on state farm land. Market gardening is increasing, to supply urban markets with vegetables. Rainfall is higher in this southern canal zone, with an annual average of 250 mm This, together with ground water seepage from the canal, allows a greater density of vegetation that can be grazed. Over the last decade, there has been outmigration by families from the desert to this canal zone, as conditions have become more arduous in the desert.

9

SHEPHERDS AND THE STATE

Effects of decollectivisation on livestock management

Christopher Lunch

Introduction

The Turkmenistan government has followed a cautious approach to reforms, very gradually relinquishing control and frequently reformulating policies. The reform process in the livestock sector has substantially changed the economic relationship between producers – the shepherds – and the state. Livestock management systems have had to be adjusted in the wake of each new policy, with consequences that are still unfolding. This chapter documents some of the impacts of policy changes between 1997 and 1999 on the livelihoods of shepherds, their methods of livestock husbandry and systems of land use.

New institutional forms of management

The Presidential Decree of March 1994 on 'Restructuring of Kolkhozes and Sovhozes and other Agricultural Enterprises' provided for the creation of associations of farms, share-holding societies and partnerships, cooperatives, associations and other forms of enterprises out of the collective and state farms. In June 1995, a decree abolished the existing large-scale agricultural enterprises and created farmers' associations (*diyhan birlishik*) in their place. Assets of the former enterprises were transferred to the associations' ownership. Land and water resources remain owned by the state, which has the right to reallocate land. Membership in the association is voluntary and members may separate with household property and household plots, but without any share of the association's enterprise or the land.

Under the new arrangements irrigated land is leased to farmers' groups, and association animals are leased to shepherd brigades. The terms and conditions of the lease are now more or less similar throughout the country. A presidential decree in December 1995 maintained the associations' responsibility for input supply, servicing of technical equipment and marketing of produce. It is found that these services are now mainly paid for by shepherds and obtained privately, although there remain cases where the association still provide such services albeit at a reduced, cost price. Farmers' associations have progressively curtailed their responsibilities in some of these spheres and individual shepherds or farmers have become increasingly responsible for managing their own expenses and obtaining necessary inputs.

Under severe economic hardship and continued pressure to follow government directives, many associations are unable to maintain their previous broad sphere of activity. Shepherds now tend to hire private vehicles when they want to take animals to the market or have water delivered to them, and have to find their own spare parts in order to maintain pumps, motorbikes and trucks. In 1998 in an attempt to prevent the collapse of some of the weaker associations the government pledged financial support for these fledgling structures and became 51 per cent shareholders.

In 1999 it was announced that shepherds who lease animals will not be allowed to keep their share of the output (R. Behnke, pers. comm. 1999). Instead they will be forced to sell their share back to the associations, at a price fixed by the associations, although this will ostensibly be based on average market prices in the region. Understandably some shepherds have expressed concern about this development, which they see are eroding their independence and ability to operate a market-oriented system. This will undoubtedly increase the power of, and the income to, the associations to the detriment of the shepherds' income.

Study areas

Institutional changes were studied in two different areas, to provide a contrast. The Yerbent peasant farmers' association, or *diyhan birlishik*, in Ashgabat *etrap*[1] (district) is one of the relatively successful new associations due to a history of close relations with Turkmen Mallory, the government agency responsible for livestock associations. Yerbent has received special treatment from the government, particularly in terms of the technical support and advice during the recent period of reform. The geographic position of Yerbent association is also advantageous, having a direct road link to Ashgabat and its livestock markets.

The second area in Gok Tepe *etrap* is composed of a number of small shepherding communities which are affiliated to different associations. The associations' administrative centres are up to 100 kms further south in the foothills of the Kopet Dag mountains alongside the Kara Kum canal. In contrast, Yerbent's administrative centre is in the same area as its production centres.

---

1 *Etrap* are equivalent to the administrative districts called *raion* under the Soviet administration. *Velayat* is the Turkmen term now applied to former *oblasts* or provinces.

The two areas reflect different aspects of Turkmenistan's reform process. On the one hand Yerbent's status as an experimental farm, with direct managerial links to Turkmen Mallory provides an insight into the direction and thought behind government efforts to restructure the livestock industry. Associations in the Gok Tepe *etrap*, on the other hand, like most of the associations in Turkmenistan have received less support and guidance throughout the reform process. A different picture has emerged in these other areas. As we shall see, the danger of focusing too much attention on a few model associations is that it has sometimes led to the creation of unrealistic reform objectives.

The leasing (*arienda*) experiment

Livestock are now leased by associations to individual shepherd brigades with an agreement that shepherds look after them correctly. In return, the shepherds must give a certain percentage of the herd/flock annual production back to the association and achieve agreed-upon production targets. Yerbent's agreement required each shepherd to achieve ninety lambs from 100 Karakul ewes. In 1998 Gok Tepe shepherds were supposed to achieve eighty-eight Saraja lambs per 100 ewes.[2] (The different sheep breeds are discussed in Chapter 11.) Any extra offspring belonged to the shepherd. However, if a shepherd did not meet the targets he had to make up the numbers with his own private sheep.

The different experimental approaches adopted in Yerbent between 1996 and 1999 traces the path taken by Turkmen Mallory in its attempts to re-mould the institutional relationship between shepherds and the state. When Yerbent became an association in 1996 the shepherds were divided into three groups depending on the quality of their rangelands and access to water. For instance those with the best rangelands and water resources received 30 per cent of the value of production from leased animals whilst those with the poorest rangelands and who had to get all their water by truck received 55 per cent since they were seen to have the highest expenses. This tiered system did not last, as it was later decided that production results were not so much dependent on the quality of the rangelands as on the ability of the shepherd.

In 1997 Yerbent association changed to a system whereby all leasing shepherds received the same agreed-upon percentage of lambs. At that stage there were significant differences between the percentages received by shepherds from one association to another. A shepherd brigade working for Yerbent association in 1997 received 80 per cent of the total yearly production, reduced in 1998 to 70 percent. Shepherds looking after association sheep in the neighbouring Gok Tepe *etrap* in 1998 received only 50 per cent of the total lambs, although they received all the wool, which was however virtually worthless at that time.

---

2 Saraja breeds are considered less hardy than Karakul and generally have a lower lambing percentage of 85–88 per cent (see Chapter 11).

In contrast to the early experiments which reflected an attempt to consider the particular environmental and geographical capabilities of different areas, the later range of leasing agreements in different associations in 1997 and 1998 did not consider the circumstances of each area. Some of the pastures used by shepherds in Gok Tepe *etrap* are up to 170 kms distant from the administrative centres, reached by sandy tracks which snake their way across the dunes. The costs involved in obtaining inputs such as fresh water are therefore very high in these sites.

The regional variation in the leasing terms in 1997 and 1998 was part of Turkmen Mallory's experimental strategy, reflecting an attempt to recognise which terms were most favourable and sustainable both to the associations and the shepherds. In Yerbent the leasing percentage was reassessed each year after the publication of the annual production figures by a board comprised of the association director, the area administrator (*hakimlik*) and representatives of Turkmen Mallory. Whilst in theory this process should be applied across the country, in most cases the percentages are decided by Turkmen Mallory without any input from individual association directors.

A large proportion of Turkmen Mallory's efforts to monitor and experiment with different approaches to reform have been carried out in Yerbent. Here it was possible to observe a number of different systems operating in parallel at any one time. By 1998 most of the state's livestock was being held by shepherds under some form of leasing contract, although some of the seventy shepherd brigades were still receiving a salary until 1999. By keeping the salary system up and running, Turkmen Mallory were effectively keeping their reform options open. The head of livestock for Yerbent association explained that the group receiving salaries were not such good shepherds and were afraid of losing money on the leasing terms. He also noted that the leasing agreement was potentially far more profitable for shepherds.

The presidential decree of 1996 stated that associations should continue to make basic inputs available to their shepherds at cost price. In reality by that stage the resources of many of the associations were too far depleted due to the chaos of economic crisis. In the Gok Tepe *etrap* since the introduction of the leasing system the associations have not been able to provide their shepherds with any inputs, including drinking water. Instead, shepherds have had to obtain inputs through private sources paying on delivery with animals or cash. In Yerbent, however, where Turkmen Mallory's presence has maintained the status quo to a greater extent, inputs such as fodder and delivered water were still available to shepherds in 1999 at cost price, albeit in limited quantities. As in Gok Tepe *etrap*, shepherds were also using private channels to acquire additional inputs.

In Yerbent, the associations' resources have been distributed among the five farms which comprise Yerbent association. These resources were then allocated to the individual shepherd brigades. A record of what was received by each herd/flock was held by the farm manager. This system is called collective leasing; the twenty or so livestock groups in each farm were considered together, the risk was shared and the accounts were settled on their behalf biannually by the farm manager and association accountant.

This situation was unique and bore very little resemblance to what was happening elsewhere in the country. In 1997 a step was taken towards officially recognising this national trend by default towards increasing self-reliance. A former head of one of the five Yerbent farms was allowed to separate from the other herds and have his own independent leasing contract. Under this arrangement the same terms and conditions applied, the only difference being that he was made responsible for keeping his own accounts and settling them directly with the association accountant. He was able to choose where to get his inputs from and how much to buy. That year the pasture was very good and he managed to economise on inputs and was reported to have made a profit of 10,000,000 manat ($2,000), enough to build himself a new house and buy a truck.

This success story encouraged other shepherds to adopt independent leasing and the system was slowly expanding in 1998. One recent convert stated:

> When you are not in the independent leasing system, payment takes place in some rather vague ways. You keep a certain number of animals, and you know the approximate expenses for the year. However, when you get your pay at the end of the year it is often very different to what you expected. They say 'these are your expenses and this is your pay'. I wanted to make everything clear, so that is why I decided to make my own leasing contract. Now things will be clear with the accountant, I keep my own record of my expenses. The old system lead to frustration, confusion and annoyance.

There was optimism among the shepherds entering into the independent leasing contracts in 1998 and as a system it seemed likely to promote better and more competitive management strategies as well as greater profits for shepherds in good years. However shepherds were also fearful of the greater risk factor and during bad years, such as very dry spring or cold winters, shepherds preferred to be part of the collective leasing to minimise loss. The administrative personnel were divided in their attitudes towards this new approach. One of their widely held concerns was that whilst this system would allow some of the stronger shepherds to prosper, the majority of younger or less experienced shepherds would suffer as their collective costs were no longer supported by the better shepherds. Therefore the association's profit margin as a whole would drop. There may also have been fears that this system would remove some of the association's power and legitimacy, making it increasingly difficult to justify the heavy administrative costs.

By 1999 those shepherds who had been so optimistic the year before had returned to the collective leasing system disillusioned and disheartened. It appeared that the administration had sought to prevent the spread of this new system and to prevent success. Independent leasing shepherds were given young or unfit flocks which required a lot of fodder and yielded less lambs. In other cases, shepherds who had asked to have an independent lease were refused permission to leave the collective system.

Leasing shepherds as a whole, whether collective or independent, were disappointed with their results and with the association's lack of transparency in its dealings towards them. Whilst previously policy had encouraged shepherds to adopt the leasing system, the settlement at the end of 1998 was a poor advertisement for leasing, as one farm manager explained:

> In my farm half of the flocks are leased. These ten flocks got better results; 150 more head. But after settlement they didn't end up with any more profit than those being paid a salary. Taxes were included in the contract but were not thoroughly explained. So shepherds were very surprised. They had expected 25–30,000,000 manat but received 7–8,000,000 manat. Much dissatisfaction flowed, even until now.

The shepherds had expected to receive their percentage in live animals. Without any consultation or warning, the administration decided that they would be given cash instead. In a climate of dramatic inflation this was clearly an unwelcome surprise, with each lamb valued at 60,000 manat which was well below the market price

At the beginning of 1999 Turkmen Mallory appeared to have reached a conclusion on the optimal leasing agreement. A standard of 50 per cent was introduced across the country with the possibility of a 5 per cent variation in exceptional circumstances. Yerbent's shepherds, for example, would receive 45 per cent of the offspring with an additional 10 per cent of dead Karakul lambs being compulsorily purchased by the association for their pelts at the (low) fixed rate of 25,000 manat. The remaining 45 per cent was for the association administration.

By 1999 there were rumours that Turkmen Mallory was considering returning to a salary system and abandoning the leasing system. In the light of the success of Turkmenistan's leasing experiment this would be unfortunate, as one of the Gok Tepe association directors explained:

> After turning to the leasing system it has been more interesting for shepherds and it is giving better results. But there is a rumour of returning to a salary system. This would be a great pity. We have just started to get the shepherds to believe in this system. I think fifty-fifty works well.

Later that year Turkmen Mallory announced that the shepherds' 50 per cent would now be paid in cash equivalent rather than in heads of animals. Lambs were to be valued at 100,000 manat by the association, far below the market average of 160,000 manat. As the 1998 settlement in Yerbent showed, these plans are likely to be very unpopular among shepherds who will be considerably worse off, especially in view of the high inflation and declining value of the manat. For the associations, on the other hand, this compromise allows them to retain the motivating elements of the leasing system whilst significantly increasing their profit margins.

All institutional structures including Turkmen Mallory have suffered as a result of the economic crises following independence. Massive cutbacks in personnel and finance are some of the reasons why Turkman Mallory has focused so much attention and expertise on a few associations such as Yerbent. There has been relatively little consultation with associations in other areas, resulting in the creation of a model association that bears little in common with associations elsewhere. The infrastructure of other associations has so far diminished in the last few years that many no longer even own a working water truck. As the associations have had to reduce their institutional responsibilities, the producers in the desert have evolved as increasingly independent shepherding entities.

It appears that Turkmen Mallory envisages other associations conforming to the Yerbent model in the future. Certainly the security of being able to request technical assistance or emergency inputs would be very useful to shepherds living in the Gok Tepe *etrap*. The ability to suspend payment until the lambs are born is also a significant advantage. However, the financial and labour costs needed to bring other associations to the Yerbent level of operation would be enormous. Even after such investment it is doubtful whether the present institutional framework would be adequate to co-ordinate these resources efficiently and economically.

Gok Tepe associations: stretched interests

Gok Tepe's association centres are located in the south along the Kara Kum canal far from their livestock producers in the desert, unlike Yerbent's location in the desert which dictates an emphasis on livestock rearing. The road and rail links to Ashgabat and other southern towns also make the canal zone an ideal location for growing vegetables, fruit, cotton and cereal crops.

A shift has taken place amongst most of these associations, away from livestock production and instead concentrating on wheat, barley and cotton with plans to further increase the areas under these crops over the next few years. This shift results from new market forces and political pressures. Ever since independence one of the Government's main priorities has been to achieve self-sufficiency in food grain production (see Chapter 11). The Gok Tepe associations have been under great pressure to achieve the grain targets and as a result a large proportion of their time and resources have been geared towards this objective while fodder production has declined.

Economic pressures are also important contributing factors to the reduced capacity of associations to serve their members. Analysis of one association's costs and returns of production in 1997 showed costs as much higher than the returns from selling the produce in markets. The extent to which accounting masks chronic inefficiencies, over- or understatement of returns and attempts to meet presidential targets is open to question.

Heads of livestock have fallen more sharply in Gok Tepe than in Yerbent. Not all of these losses were suffered after independence. In 1978 the collective farms (*kolkhoz*) of Gok Tepe were restructured and each farm became specialised in one line of production. This attempt demonstrates that even before the recent chaos brought on by political and economic reform it was felt that collective farms might perform better if their interests were not stretched between livestock and cultivation. Shepherds say that specialisation was an improvement and meant that they needed to use less fodder and were able to increase production. Despite this, the policy was reversed a few years later. When the animals were returned to their original *kolkhozes* there were often fewer animals than before. For example, Association A had 12,000 sheep prior to specialisation but only 2,000 sheep and 100 camels were returned. The extent of loss suffered since independence differs from one association to another; the most fortunate associations may now have half the number of sheep they had ten years ago whilst others such as Association A now have so few sheep left (approx 500) that it is no longer worthwhile keeping them in the desert. One association staff member said:

> There have been so many reforms and reorganisations; animals have been taken away and then given back. When the time of self payment began and the currency changed [1994], we had to pay salaries and the maintenance of equipment with our camels and sheep.

The associations' economic hardship, the sheer distance from the administrative centre to the pastures (up to 170 km) coupled with market pressures and government incentives are all contributing to the isolation and abandonment of shepherds remaining in the desert. The reduction in the number of flocks kept by the different associations has also had a negative impact on these shepherding communities. A senior staff member estimated that from his association alone at least thirty former shepherds had lost their livelihoods over the last six years. Many of these families have now abandoned shepherding and have moved south to the association centres along the canal, in order to lease small plots of land for cultivation.

Table 9.1 shows that Association A experienced a dramatic drop in heads of sheep between 1996 and 1997. The former state farms became associations in 1996 and the stresses of reform can be seen close up here as animal productivity plummets.

Sheep production dropped by 50 per cent in this association in one year, while 15 per cent of adult sheep were reported to have died. Not all these animals would have died from natural causes; most were probably sold or exchanged in order to cover the association's running costs. The results for cattle and pig productivity were also far below target. Under favourable conditions every 100 cows were expected to have eighty calves, but in 1996 and 1997 only a quarter of these were obtained. These poor results probably reflect the disorder brought about by a collapsed infrastructure and institutional reorganisation. Pressures to pay immediate costs with animals may have been disguised as natural mortality. However,

Table 9.1 Planned and actual livestock production figures for Association A between 1996 and 1997

Livestock	Total heads		Actual offspring per 100 mothers		Planned offspring per 100 mothers	No. adult stock died	
	1996	1997	1996	1997	Plan	1996	1997
Sheep	677	573	43	88	80	103	30
Camels	72	80	5	40	40	0	0
Cattle	316	325	24	29	80	16	6
Pigs	156	141	344	236	1,200	34	19

observers agree that with the removal of state support, fodder was very scarce and distribution channels were disrupted, which had the greatest impact on stall-fed animals (cattle and pigs) in the canal zone.

A bleak picture emerges from these figures. The challenge faced by association directors is to adapt their management strategies to the shifting demands of this transitional period. The old collective system grew out of a political and economic climate which has little in common with present market demands and the new organisational structures. Each association is having to find its own way through the uncertainty of reform. Unfortunately, during this time of experiment the costs are often high.

New demands squeezing the shepherds

The decline in sheep numbers among Gok Tepe associations is said to have tailed off between 1997 and 1998. In 1999 a national target to rebuild herd sizes was set for associations to increase their heads of livestock by 5 per cent each year until the year 2005. This demand has put considerable pressure on the already failing associations, whose staff were trying to find a way to achieve this challenging target. Associations now had to keep older sheep and slaughter fewer for meat, adding to their financial strain as income from meat sales is reduced and higher input costs are incurred to maintain older sheep (see Chapter 11). One association had persuaded its shepherds to exchange their share of female lambs with male lambs belonging to the association, so that the association would have more ewes and thereby increase production. The Soviet-style hierarchical structure remains and if a association underperforms or the etrap as a whole does not achieve acceptable results then the association director or district administrator (*hakim*) are held personally responsible and may lose their positions.

Turkmen Mallory's next proposal to pay shepherds their percentage in cash equivalents rather than in heads of animals provided the associations with some extra income and a better chance of achieving the government targets. Therefore it was the shepherds who ultimately had to shoulder the consequences of this latest squeeze. From the shepherds' point of view this was an unwelcome development, not least because it signifies a return to a system based on payment from the association, one in which they had very little confidence.

> We did not like the pre-leasing system. Maybe the *kolkhoz* didn't have enough money to pay us. We never received our salaries. When the leasing started I submitted my livestock to the association because I did not know what would happen. I wanted to see the results. Now I think I will join the leasing when it is a good year. We realise now that the leasing is good if you work hard. None of the shepherds in our village have faced any loss.

The latest adjustments to the leasing system will undoubtably make it less attractive to many families. In the desert villages closest to the association centres the lure of the canal zone is particularly strong. Already these communities have been torn apart as many of the relatives and neighbours of those remaining families have moved away. Desert families are now balancing their love for their animals and the environment in which they have always lived, against the relative ease of life in the canal zone where there is electricity, gas, schools for their children and the possibility of leasing land for cultivation. The less attractive leasing arrangements could be the last straw forcing yet more shepherds to leave the desert.

In trying to rebuild the country's livestock, these latest reforms may escalate the very problem they are trying to address, by increasing the level of out-migration from desert villages. As one shepherd explained, this could have irreversible consequences for the future of Turkmenistan's livestock industry.

> If a big family like ours looking after one association flock moves away from the desert then this is one flock of sheep which is lost to the nation. Once people move to the south they never come back to look after sheep.

Impact of reforms on associations

The 1999 government directives did not address some of the more fundamental obstacles to reform, focusing instead on symptoms rather than causes. The financial pit from which the associations seem unable to emerge and their failure to establish themselves as successful entities is not due to any shortcomings of the leasing system. Indeed, the 50 per cent the associations were to receive from leasing shepherds was very generous, considering how little the associations provided to shepherds. It is rather the result of their Soviet heritage and the idiosyncrasies of the *kolkhoz* structure on which they have been built. Certain aspects of this heritage are making it very difficult for the new associations to adapt to economic demands and are obstacles to successful and rapid reform.

Many of the inherited features of these associations, such as the weighty administration, the Soviet-style accounting systems and the hierarchical structure sit awkwardly with the new demands of a market economy. The hierarchical system and job insecurity breeds misinformation and distortion. As in the Soviet period, statistics are adjusted and information is selectively handed up the power ladder. The association administration is pulled in many different directions and the directors have to try and carry out all the different orders thrown at them by their superiors. A huge amount of their energy and capabilities are spent simply keeping their heads above water and trying to please rather than attempting to tackle the real challenges of reform. The rapid turnover in personnel is itself affecting the speed and quality of reform, since experience and knowledge about the new economic and social realities cannot be built up.

These remnants of the past are the cause of great frustration for those association directors who wish to succeed. One association director complained how up to 50 per cent of the association's meat profits were consumed by its workers either for free or at half the market value. For example during harvest time all the workers have to be fed, and as he noted:

> This is the old system; we have to provide it or we would be sacked. It is just sucking us dry. If it wasn't for this then maybe we would achieve the 20 per cent increase of livestock.

The direction of reform over the period 1996–9 shows that Turkmen Mallory places the associations at the centre of their vision for the future of the livestock industry. Whenever it has appeared that developments might make these associations obsolete, they have always received some sort of boost from the government, with the adjustment to the leasing agreement being the latest example by 1999. The collapse of the Soviet Union took the heart out of the collective farm system. Turkmen Mallory's aims to put life back into these institutions, bringing them to a level where they can provide shepherds with low cost inputs and assistance, would require enormous financial investment. For these institutions to be viable and sustainable they would also require some dramatic structural changes. So far experiments and reforms have focused on allocation of economic returns and have not addressed the structure of the institutions themselves.

Turkmen Mallory's lack of resources has meant that there has been little or no monitoring of the results of reform on producers. Instead Turkmen Mallory continues to work within an environment of misinformation and distortion. While shepherds are organising themselves into new systems of management, the formal institutional structure presents an obstacle to the progress and change taking place in the desert.

New forms of livestock ownership and management

During the Soviet period shepherds were restricted in the number of livestock they could own privately. The government has now removed taxes levied against an individual's private livestock in order to encourage the growth of this sector. Recent official statistics suggest that the fall in state-owned livestock has been matched by an increase in privately held livestock (see Chapter 11). Whilst there has been considerable increase in private ownership of livestock, shepherds in the southern part of the Gok Tepe desert pastures had witnessed an overall drop in animal numbers since independence. Initial investigations suggest that in the more remote northern pastures there have been less significant reductions in total animal numbers (R. Behnke, pers. comm. 1999).

Private flocks and herds are composed of animals belonging to different members of an extended family. Traditionally a father gives some of his livestock to his sons when they marry, and in some cases married daughters also have a share in the private herd. Most desert families also keep between five and ten private milking camels. Privately owned livestock is currently being managed in one of three ways:

Mixed association and private herds

Most association shepherds have been able to increase their own private herds under the leasing agreements. These animals are herded together with leased animals and treated with equal care. Three or four individuals are needed in order to manage the minimum-sized association flock of approximately 300 sheep. At present the number of association flocks available to shepherds living in the Gok Tepe area is limited and are generally held by families with adult sons or groups of brothers. If successful, government targets to increase the number of animals held in the state sector could lead to the creation of more association flocks in the near future. However, amendments to the leasing terms in order to achieve an increase in state-owned livestock will inevitably slow the growth of private flocks.

Entirely private herds

In the Gok Tepe desert region families have depended on shepherding for the state farms over a long period. For this reason it is relatively rare to come across families now surviving with only their private livestock. Most privatised shepherds have either had to move away or become shepherds for other owners, as described below. There are exceptional cases, but mostly these families either own a truck from which they can receive income or are in some way connected to a larger extended family who may in turn have an association or hired flock.

Absentee herding

The Turkmen word *chekene* is used to describe the increasingly important management strategy of absentee herding. Families who are unable to look after their private stock or have too few animals to make it worthwhile, can pay a shepherd to look after them. *Chekene* shepherds and absentee owners are usually friends or kin since the arrangement depends upon mutual trust. The average fee for this service in 1998 was 2,000–3,000 manat per sheep per month and up to 15,000 manat per camel per month. These animals are kept together with the shepherd's own private flock but their offspring belongs to the absentee owner. Owners may choose to slaughter or sell the offspring or keep them with the shepherd in order to increase the owners' holdings. In some cases individuals may choose to increase their flocks by buying animals from the *chekene* shepherd. Absentee herding existed before and during the Soviet period but has become increasingly widespread since independence. As an adaptive management strategy it enables those families remaining in the desert to maintain a livelihood and provides them with an alternative to keeping an association flock.

Two different types of *chekene* shepherd can be identified, according to the circumstances in different locations. In Yerbent, due to proximity to the Dashauz–Ashgabat highway, there are a number of opportunities for alternative skilled and unskilled work. In Yerbent village there are approximately 300 families of which only twenty look after association flocks. Others work in the administration or on the association's irrigated fields. Many families in this village have at least one member bringing in a wage as a driver, builder, teacher or guard, etc. An increasing number of families are setting up tiny shops along the road and there are also some small-scale entrepreneurs involved in trading and selling goods. Everyone has at least a few head of sheep and one or two camels. Therefore there is considerable demand for *chekene* shepherds to graze these animals. Often families take their animals back in the winter, to be stall-fed in their backyards. The same pattern can be observed in the villages situated along the Kara Kum canal in the southern irrigated region of Gok Tepe *etrap*.

In more remote desert villages of the Yerbent region, as with the Gok Tepe desert communities, nearly all those people who are not directly engaged in livestock-rearing have moved away. This has led to the rise of a different form of *chekene* herding, as families migrating south left their private animals in the care of neighbours, friends or relatives who stayed behind. As these areas are depopulated, larger numbers of livestock are being concentrated in the hands of the remaining stronger shepherd families. These families have a sufficient livestock base, a number of adult sons and enough influence and contacts to ensure access to key resources such as trucks, water pumps, spare parts and fodder.

The *chekene* system is reinforced by customary patterns of urban–rural exchange. Sometimes agreements are made whereby the families which move to the irrigated areas are responsible to the desert-based families for obtaining cheap fodder, providing a truck to take animals to the market or delivering water, etc. These networks of reciprocity are very important during this reform period. Getting hold of inputs is often time-consuming and expensive and any savings are of great value.

The demand for long-distance *chekene* shepherds is growing as more people decide to hold their private flocks in the more remote regions where there is less chance of theft and disease and where there is less competition for pastures and water resources. Livestock are regarded as an investment as well as a form of food security, even by Turkmen living in the cities. When the oil starts to flow and more individuals in the city begin to accumulate wealth we may see the appearance of larger *chekene* flocks and herds in these remote desert villages.

Changing patterns of resource use

In the Soviet collective farms, each flock/herd was allocated specific areas of pasture to graze and access to specific water resources. This defined territory of land with its own water supply is called an *oiye*. This system still functions in Yerbent as the association remains under relatively strong centralised control. In the Gok Tepe area, however, each flock can now be moved more freely in search of the best land and water. Shepherds now have greater responsibility in the way they manage the animals in their care. Reforms since independence have created a new set opportunities and limitations for shepherds, relating to financial considerations, questions of access as well as natural and climatic factors. Individual management strategies must now hinge on a complex interplay of all these factors and are explored in this section.

In this arid environment water is a key factor determining pasture use and influencing decisions on when or where to move. Decollectivisation has significantly changed the way in which water resources are managed and maintained. The introduction of the leasing system means that shepherds are now much more aware of the financial implications of different management strategies. These factors have initiated a process in which shepherds reassess the available choices, a process which is continuing. Some of the ways in which these changes may affect the area's future capacity to hold livestock will be considered below.

There are five main water resources available to Gok Tepe shepherds in the southern desert region: underground wells, pipelines, delivered water, run-off rainwater and, lastly, drainage water from the Kara Kum canal.

Wells

Wells are scattered throughout the territory, in villages, hamlets and in the pastures. Whilst the ground water level is consistent throughout the territory, its quality is widely variable. Most villages are situated around *chyrla*-type wells (see Chapter 10) which are always found in connection with natural depressions known as *takyrs*. The quality of these wells varies seasonally; sweeter in the spring but becoming bitter as the underground lens of water reduces. After a dry spring the water in most *chyrla* wells becomes unusable. Even in a good year the water becomes bitter as summer progresses, making the animals sick and causing diarrhoea.

Pipelines

In Yerbent association many shepherds are provided with water by the pipeline which stretches from the Kara Kum canal into the desert. Smaller pipes branch off the main pipeline and carry water to some of the more remote villages within the territory of Yerbent association. In the past a plastic pipeline carried water pumped from Bahardoc village, belonging to Yerbent association, west into a pasture area traditionally used by Gok Tepe shepherds during the winter. The pipes would continually burst and were not properly maintained. The source was eventually cut off in 1996/7. Well water in this area of Gok Tepe is very bitter and the pipeline provided an important source of fresh water. The collapse of this pipeline link has discouraged many of the southern Gok Tepe shepherds from taking their herds to these traditional winter pastures.

Water brought in by truck

Heavy duty water tankers became a lifeline for the shepherds living in these arid environments in the Soviet period. Tankers were used frequently but now the decision to order water involves careful consideration as the costs are high. Water tankers may be owned privately or by an association and are usually equipped with pumps enabling them to fill up from a variety of sources depending on the needs of the customer. Water for human consumption is taken from mountain streams in the foothills of the Kopet Dag mountains. A delivery of four tons of this water costs 150,000 manat or a young sheep. A submerged reservoir tank or *sardop* outside each house holds approximately eight tons of water, enough for one family for a whole year. In summer when the well water is bitter this water is used for tea and cooking, while in winter when rains sweeten the wells it is only used for tea.

Water for animal consumption is also brought in by tanker. In spring, shepherds are unable to move their pregnant ewes and young lambs so if there is no rainwater available the tankers must be filled with imported water. For the southern desert villages this water is usually taken from nearby channels carrying waste water from the irrigated zone out into the desert. One shepherd explained that to transport four tons of this drainage water twenty kilometres to his spring lambing area cost 35,000 manat. People who own cows in the desert also import drainage water as cattle cannot tolerate the bitter well water.

Often one part imported water is mixed with two parts bitter well water to reduce costs. This technique is especially used in the remote northern pastures where well water is very bitter and the costs of transporting fresh water are high. The insecurity and costs involved in supplying flocks with water in the northern pastures was one of the major factors discouraging new private shepherds from making the traditional winter migration (described below). One shepherd calculated that even if he mixed bitter and fresh water the costs of watering his herd of 500 sheep over five months in the winter pastures amounted to 20 per cent of his total expenses for the year. In Yerbent the association still supplies its shepherds with water when they need it and delivers it to shepherds in the winter pastures. Whilst shepherds have to pay the association for this service, the settlement of their accounts does not take place until the new seasons' lambs are born, thus providing a form of deferred payment. This has to a large degree enabled Yerbent association shepherds to continue their winter migration.

Run-off rainwater

Throughout this desert region, dotted amongst the sand dunes are flat areas of clay rich sand known as *takyrs*. These can be anything from a few metres square to the size of three football pitches. Natural pools form on *takyrs* during the winter and spring rains. In some cases man-made basins (*kaks*) have been dug into the lowest point of the *takyr*. Rainwater flowing across the *takyr* is channelled into these basins and used to water animals. Reservoir tanks (*sardops*) have sometimes been submerged in the ground alongside the basins, or else free-standing metal water tankers have been placed nearby. These tanks are filled naturally or by pumping water from the basin and significantly reduce the water lost through seepage and evaporation. Water held in this way could last a shepherd's livestock for up to two months depending on the size of the reservoir tank.

Kaks need to be redug every two years and any plants growing in or near the *kak* should be removed in order to maximise the amount of water they can hold. Only a few shepherds were encountered who still regularly maintained and used their *kaks* and many *kaks* are now grown over. One explanation is that many shepherds started relying on water deliveries during the Soviet period and lost the habit of using *kaks*. Another possible reason is that spring rains have been poor and unreliable over the last ten years and under these conditions *kaks* are relatively useless.

Now that financial considerations have entered into the management calculations of individual shepherds it is likely that *kaks* will become more important. *Kaks* provide shepherds greater flexibility of movement since it is not necessary to return to the wells for water, allowing use of more remote pastures. Shepherds maintained that rainwater was healthier for sheep than the salty water found in most wells, since it was more easily digested and enabled sheep to eat the salty-tasting plants which are considered medicinal. Some shepherds recognise the potential benefits of using kaks but did not have any historic rights of use over one. To dig a big *kak* requires a tractor and there was a reluctance to dig smaller *kaks* by hand as it was not felt to be worthwhile.

Takers also require regular maintenance by local users. If plants are allowed to grow they act as barriers to the sand blowing across these large flat areas, which eventually leads to the creation of sand dunes on the *takyrs*. Cars travelling across wet *takyrs* break the surface crust and encourage plant growth (Batyr Mamedov, pers. comm. 1998). Once again the result is more rapid deterioration of the water resource.

Often *takyrs* are no longer properly maintained and only one or two older individuals still take time to clear the plants. In some villages with strong local leaders the communities have taken action. For instance in one village families had co-operated to buy some wire and prevent cars from driving across their *takyr*. In other communities a third party may be necessary to pull the various figures of authority together and initiate discussion on important issues which affect the whole community.

Drainage water

All waste water from the farming communities along the Kara Kum canal in the south is directed along channels into the desert region. This consists of run-off water, human sewage, water used to wash leached salty minerals off the fields and excess water from irrigation. Although these channels were built approximately fifteen years ago the volume of water they carry has increased dramatically in recent years. This is partly due to agricultural reforms and the arrival of inexperienced farmers onto the land, resulting in badly managed and poorly maintained irrigation systems (O'Hara 1997).

Local shepherds have mixed feelings towards this drainage water. As the water flows across the desert it becomes less bitter and is used by local communities as an important source of water for animals and an alternative to bitter well water in the summer. The area is well-known for the quantities of *yandak* (camel thorn) which shepherds cut from June until October as winter feed for their animals. *Yandak* is especially prized now that cultivated fodder is so expensive to buy. The drainage zone has therefore become an important feature in the seasonal management of herds based in southern Gok Tepe desert pastures.

There are also some very tangible negative impacts of this drainage water. Reeds that have grown up around the drainage channels provide shelter to many jackals and foxes. These have caused such losses to the sheep population in nearby villages that only camels are now kept locally. The water often infects animals with liver worms and some shepherds try not to use the drainage water for this reason. Many shepherds have few other options and have resigned themselves to treating their animals regularly with medicine. This adds extra costs and puts a strain on the condition of the animals, sometimes causing mortality.

In some areas pools have formed and the stagnant water has become a breeding ground for mosquitoes and gadflies. In villages more than 30 kms away from a drainage channel camels were contracting diseases as a result of these gadflies. Under these conditions it became impossible to keep large herds of camels in the surrounding area. Between 1985 and 1992 all *kolkhoz* camel herds were moved from the southern desert pastures to more remote northern villages. Families continue to keep private camels but these must be given an injection twice a year to protect them. The impure water has polluted wells in villages immediately adjacent to the channels and the fluctuating ground water levels have eroded the base of wells in the surrounding region, causing them to collapse in some cases. Drainage channels are no longer maintained and regularly overflow onto surrounding land. This results in the creation of more stagnant pools and often desert tracks crossing over the drainage channels are flooded and impassable.

Despite its usefulness the drainage water has polluted the local environment for miles around and has already significantly reduced the area's capacity to keep livestock.

Well management and maintenance: in theory and practice

A discrepancy exists between the official status of wells and their actual usage. Officially, wells remain under the control of the individual former state farms (now associations) which built them and which remain responsible for the maintenance and management of these wells. Most villages have some wells which belong to associations and others which are reserved for communal use. In Yerbent association the distinction is often very clear, for instance in the village of Bori two of the five wells were equipped with a pump and were fenced off for use only by association shepherds. In the Gok Tepe *etrap* the associations' reduced capacity has diminished their ability to apply central administrative control. Officially a private or *chekene* shepherd wishing to use a association well for an extended period of time, such as over winter, would be expected to pay the association a sum of approximately two sheep per month. However, private shepherds are avoiding these costs either by keeping a low profile or by teaming up with friends or relatives who look after association herds. These new alliances are increasingly common and are sometimes solidified by arranged marriages between the co-operating families, demonstrating the key role family relationships are playing in the evolving management strategies. Since there are no satisfactory means by which associations can protect their resources a laissez-faire attitude prevails.

At present resources are used and managed from a local perspective and the needs of a neighbour are more important than loyalty to an association. Access is justified on the basis of historical use patterns and a shepherd may claim the right to use a certain well or *kak* because his forefathers did. Such claims may be irrespective of whether he is an association member, private or *chekene* shepherd.

> Getting access to water can be solved by hiring a local man to look after your animals. You pay 2,000 manat per head per month and they sort out the water. A local man will not have any problem finding water even when the animals are not all his; it is considered his enterprise.

This statement made by a local shepherd indicates that in this community at least, the right of *chekene* shepherds to use the village well has not been disputed. At present few conflicts arise over questions of water access and when they do, they are resolved by the elders. In most cases codes of honour are sufficient to ensure a system which is mainly self-regulating. To deny another shepherd access to one's water or to abuse another person's supply is seen as shameful by the rest of the community.

But misunderstandings and minor conflicts do occur. One association shepherd complained how the huge *kak* and *sardop* to which only he and two other association shepherds had official access, was regularly used by local village residents taking water for their cows. Under present conditions the association shepherds felt unable to stop this. In larger villages where competition for resources is greater, more conflicts do occur.

Given the decline in usable water points, increases in flock sizes – whether they occur in the state sector in line with government targets, or in the private sector – would increase local competition for resources. Competition could destabilise the existing system. The example from Yerbent suggests that, under such conditions, distinctions between association and private/*chekene* flocks are likely to become more defined and stricter controls will need to be introduced. Whether these controls will be imposed by local users or by association administrations is unclear. Present trends indicate an increasing level of self-governance and control at the local level. However these trends are largely a result of inaction on the part of the association centres rather than a reflection of any explicit policy.

The associations' passive role regarding management of wells is matched by their inability to maintain them. This has significant implications for the areas' ability to sustain herds in the future (discussed in Chapter 10). Whilst the local people are content to be given greater freedom in their use of available wells, maintenance is still regarded as something which should be carried out by the associations. The result is that neither the local communities nor the associations are maintaining wells. The area surrounding the wells should be dug out every five years by a tractor digger. In most villages this has not been done for up to fifteen years. If wells are not dug out, the quantity of rainwater collected near the wells and filtering down is reduced. In addition, a hard surface crust develops which inhibits the infiltration of rainwater. In these cases, livestock are not able to rely on the wells for so long. In some villages, rainwater has overflowed the well area and damaged nearby houses. Although residents realise that the associations will not dig the wells out for them, the alternative of hiring a tractor for one day is said to cost too much, even if divided between ten families. If the association or local government could provide a tractor at a subsidised price the shepherds would be happy to contribute to the costs.

In some of the villages in the southern Gok Tepe area, wells are under threat of collapse as a result of rising ground water levels from the Kara Kum canal. General lack of maintenance can also lead to wells collapsing, which is already affecting local people. Shepherd families have had to move away from villages when their wells collapsed, or to rely more heavily on water brought in by truck. Wells in more remote pastures are particularly vulnerable, suffering in recent years from neglect and under-use as a result of depleted stock numbers and reduced mobility. A reduction in the number of useable wells in these areas will limit access to grazing land and the capacity to increase flock sizes in future. If this lack of maintenance is happening on a national scale, there will be serious consequences for the future recovery and growth of Turkmenistan's livestock industry.

Changes in seasonal patterns of movements

Shepherds divide the desert region into three distinct pasture areas. Each pasture type has a definite geographical area and preferred seasonal use (Table 9.2).

Table 9.2 Turkmen and Russian terms for the seasons

Season	Turkmen	Russian	Number of days	Duration
winter	*gysh*	*zima*	90	1 Dec.–28 Feb.
spring	*yaz*	*visna*	76	1 March–15 May
summer	*tomus*	*letta*	123	16 May–15 Sept.
autumn	*guyz*	*osin*	76	16 Sept.–30 Nov.

Ala (patchwork)

This term describes the varied landscape in the southeastern part of Gok Tepe *etrap*. Here, sand dunes of differing height are interspersed with flat *takyrs*. Three different microclimates are represented in this area; in the *takyrs* grow salt-rich succulents, the low dunes are characterised by small bushes and scrub, and the high dunes are covered in tall grasses and trees. This area is used in spring when vegetation is abundant and if rainwater collects on the *takyrs* and *kaks* the young lambs can be watered with the fresh water. As spring progresses and vegetation is reduced, flocks become difficult to control as they spread out in search of forage. At this point (May) flocks are moved south west.

Bitew

This area is vast and flat and therefore flocks are easier to control. In the summer there are no ticks in this area and the well water is sweeter than elsewhere. Flocks are watered regularly either at three sweet wells or at the drainage channels.

Tum (wilderness)

Until the recent past, most shepherds would move north to more remote desert pastures in autumn and winter. These areas are unsuitable to graze at other times of year because of ticks and a lack of sweet water. In winter animals need watering less regularly and are able to exploit the untouched vegetation. Whilst wells are available in *tum*, they are very bitter and the water must be mixed with delivered water to make it more palatable.

At present, seasonal movements rarely correspond to the above pattern. As has been discussed, the size of flock, form of ownership and the resources available to individual families are crucial factors influencing individual management strategies. The new strategies adopted by shepherds are also conditioned by Turkmenistan's variable climate and rainfall.

In Gok Tepe many shepherds move to the southern irrigated area in the autumn before taking their flocks to the northern winter pastures. Sheep graze on the harvested wheat, alfalfa and grapevine fields. However, some shepherds avoid moving their sheep to the irrigated areas as the animals often contract diseases. One shepherd had calculated that worms contracted in the south had cost him an extra 500,000 manat for medicines and 200,000 manat for additional fodder. The leasing system had forced him to look more closely at his expenses and returns and had led him to the conclusion that this was an uneconomical migration. In addition, redistribution of cultivated land means that shepherds now have to pay to graze their sheep on the harvested fields. All the same, in autumn 1999 the northern pastures were in such poor condition as a result of low rainfall in the spring that nearly all the shepherds in the southern desert regions took their sheep to the irrigated fields in order to fatten them up in preparation for what promised to be a difficult winter.

The traditional northern winter movement is also being re-evaluated. Shepherds with smaller private or *chekene* herds now rarely undertake this movement as they have found that it is actually more economical to keep their herds near the village and feed them additional fodder. Even larger mixed association and private shepherds try to avoid moving so far from their homes. In the past, state farm trucks would help shepherds transport tents and basic equipment needed for this extended stay in the desert. Now some families would have to hire a truck or use up precious fuel resources making the journey back and forth on motorbikes in order to keep the shepherds supplied with food. In 1997 only two association herds went to the northern desert and they were not supplied with water by the association. The next year, 1998, was a good year and the pastures near the villages were able to sustain herds through the winter.

All shepherds, whether association, private or *chekene*, are now relying more heavily on pastures near to the villages. These pastures are fully grazed before individuals decide to move further away. This is increasing the risk of land degradation, as discussed in Chapter 10. There is an emerging competition for accessible grazing land, which can be observed in Yerbent, for example, where the association has remained a strong local force and animal numbers have not decreased to such an extent. Particularly in the larger villages where everybody has at least five sheep and some individuals have up to sixty private goats and sheep, association flocks occupy most of the available pastures further away from the villages, apart from some very remote regions. Private shepherds based in these bigger villages are therefore effectively locked in and limited to grazing the often severely degraded land immediately around the village. One individual described how in the summer he grazed his animals within 5–6 km of the village 'as far as you can walk and return in one day'. The area he grazed was shared by four other flocks. Whilst there was enough vegetation in good years, in years of poor rainfall tensions arose and there were sometimes arguments over territory. In winter, private shepherds were able to graze land vacated by association shepherds as these moved north to fresh pastures.

Shepherds having both association and private animals are more able to move as their flocks are bigger and economies of scale justify movement. Under present conditions, in the event of a drought year it is the smaller private / *chekene* shepherds who will suffer the most, being unable to move to other pastures. The association in Yerbent has continued to provide inputs and technical assistance to employed shepherds. This support has enabled them to continue their northern winter migrations and reduced animal losses suffered in periods of low rainfall or harsh winters. The journey back from the winter pastures to the villages is often long and hard. In the past the *sovkhoz* would assist by transporting weak sheep by truck or supplying the shepherds with extra fodder. Yerbent association still provides a degree of support in this situation, although there is rarely any fodder left by spring. Gok Tepe shepherds, on the other hand, no longer receive any assistance from their associations.

Yerbent leasing shepherds undertake this migration of their own free will and pay for the extra expenses incurred themselves. The fact that their sheep will be healthier grazing the untouched winter pastures would not be enough to make the shepherds move. They also need a degree of financial security. The knowledge that help from the Yerbent association would be available if needed and the ability for Yerbent association shepherds to suspend payment for these services until the lambs are born is what continues to make winter movement possible.

So far there have been no really cold winters or extended droughts since the formation of the associations. Under the new arrangements, with shepherds and farmers becoming increasingly independent, disastrous climatic conditions could cause massive livestock losses. In the Soviet period dramatic rescue operations were sometimes carried out. For instance, in 1986 thousands of sheep were transported by trucks from western Turkmenistan to better pastures in the east in order to prevent massive livestock losses due to severe weather (O. Hodjakov, pers. comm.). Under such unpredictable climatic conditions and at this stage in the reform process a degree of state support would be advantageous.

Future challenges and opportunities

The Turkmen government has not rapidly and completely dissolved the institutional structures inherited from the Soviet agricultural system, in contrast to Kazakstan. Evidence presented here suggests that while this may have been wise, reforms now need to focus on institutional restructuring in order to transform the relationship between shepherds and associations. It would not be realistic or advisable to try to replicate the Yerbent association model elsewhere in the country. But associations could play a useful role in redeveloping Turkmenistan's livestock industry if they were limited to making technical services available to producers on a cost-recovery basis as in Yerbent.

Producers' adaptations to the new economic and institutional conditions have been both rapid and resourceful, although they are being limited by certain factors. Without support for shepherds to move into more remote pastures, which are at present under-used, land around the villages will continue to degrade. If communities could rent water trucks from the associations they could supply their herds in the winter pastures. The reduction in private costs and increased security would encourage more shepherds to undertake this migration. Likewise, repair and maintenance of water points on a cost-sharing basis between shepherds and the state would open up more pastures and reduce localised grazing pressure.

Communities have been torn apart by the collapse of the Soviet Union. Schools have closed down and there is increasing livestock theft. Living conditions in these marginal environments have deteriorated and high levels of out-migration have taken the hearts out of many villages. To varying degrees, a new individualistic attitude has emerged with the formation of small closed groups of relatives. In Kazakstan, which has gone further along the path towards privatisation, economic differentiation has become sharper and an atmosphere of mistrust and jealousy is found in rural communities. In Turkmenistan, whilst close kin have strengthened their links, there remains a greater degree of solidarity and co-operation between neighbouring families in the desert villages.

To replace the state support which has been withdrawn, communities will need to strengthen their resource base and levels of cooperation. Support for the trend towards greater local control would help communities to take greater responsibility for their immediate environment and resources. For instance, if a tractor digger was made available at a reduced cost or on credit, villagers would be able to maintain their hydrological structures and increase water capacity. If responsibility for maintenance and management of resources remains ambiguous and shepherds continue to wait for associations to come and solve their problems, the infrastructure will continue to decay and yet more producers will be forced to move away. This will further threaten the long-term sustainability of livestock rearing in the desert.

In order to avoid future conflicts and build local capacity for fair and sustainable management of local resources, discussions should be initiated between the different local users. Resource management is tied up with many other local issues and power relations and may not always be resolved in the interests of the smaller producers. Information exchange between different communities in structurally similar situations and guidance by third parties trained in facilitation methods could ease this process. Now is the best time to discuss these issues while competition for resources is still low but may escalate if and when livestock numbers increase.

The challenges faced by the livestock industry and individual producers are great, and continually changing. The shepherds and association administrations have been learning to rear livestock under free market conditions, with some positive results. Although the economic climate remains unsettled, organisational structures must continue to be developed and refined, community bases must be strengthened, and infrastructure which is in tune with the new economic demands must be developed in these production centres.

10

THE LIMITS OF THE LAND

Pasture and water conditions

Khodja Khanchaev, Carol Kerven and Iain A. Wright

Introduction

Livestock management in the arid Kara Kum desert depends upon access to water. Livestock can only use pastures if water of tolerable quality and in sufficient quantity is available. The sheep and goats mainly kept in the Kara Kum can graze only within a limited radius of water. This means that when acceptable water supplies are unevenly distributed, grazing intensity must likewise be uneven, with consequences for the vegetation condition. Pastures close to a water point will be more heavily grazed, while pastures without access to water will be grazed relatively lightly, in some instances only by camels, which require water far less frequently.

The efficient utilisation of Turkmenistan's desert pastures is closely connected with the placement of man-made constructions for watering livestock. Unfortunately, at present in Turkmenistan millions of hectares of pastures remain without livestock drinking water and consequently the forage of these resources is not used.

This chapter assesses the condition of vegetation and water supplies used by livestock under current management systems. It is found that changes associated with reform of the state farms may be increasing the risk of localised land degradation.

The uneven supply of livestock drinking water and its seasonal availability is the main obstacle in applying more appropriate grazing management. The unsystematic pasturing of smallstock due to water limitations decreases livestock productivity and increases pasture degradation. Due to these two problems there is an urgent need to increase and improve the supply of livestock drinking water in the desert pastures. As overgrazing is already evident in some areas, in future it will be necessary to rehabilitate some areas that have lost their natural fodder base.

Methods of study

Research was carried out in the central Kara Kum desert in 1998, covering Yerbent farm association in Ashgabat *etrap* (district). Data was collected along a regional

194

transect of 150 km length from north to south (starting at Bahardok approximately 100 km north of Ashgabat) and 25 km width from east to west, on pasture productivity by season, species composition of vegetation and problems of water supply.

A total of five transects (200 × 4 m) along the Yerbent regional transect were selected for detailed seasonal measurements. Within each transect ten quadrats (1 × 1 m) were selected. Observations were made in spring, summer and autumn 1998. For shrubs and dwarf shrubs measurements were made on the transect, but for grasses and other ephemeral species measurements were made on the small quadrats. Measurements were taken on vegetation species, abundance of species, percentage of ground covered by vegetation, total dry mass of vegetation per hectare, dry mass of the portion available to grazing/browsing livestock. For shrubs, the total number of species present and the number of plants of each species was recorded. For each species, all leaves were collected from a sample of the shrubs, dried and weighed. The total dry mass of leaves of each species was then calculated by multiplying by the number of plants of each species. For grasses and other ephemeral species, all the vegetation within the 1 × 1 m quadrats was cut to ground level, separated into species and dried. Calculation of the proportion of the total dry matter that was available to livestock was based on Nechaeva (1985). The fodder value of the available portion was then calculated from chemical analysis using the methods of Nikolayev (1980). The fodder value was expressed in fodder units. One fodder unit supplies the same amount of energy to a fattening animal as 1 kg of oat grain. A fuller description is given in Appendix I, page 257.

A second regional transect of 200 km north to south and 20–5 km east to west was investigated in Gok Tepe district during 1999. This regional transect covered four ecological zones according to soils, precipitation and pasture characteristics. The zones were: northern desert, central desert, canal zone and foothills of the Kopet Dag mountains. It transpired that these zones had been distinguished and named by local pastoralists a long time ago, and that there was almost complete correspondence between our subdivisions and those made by users of the transect area. Similar measurements were taken on pasture productivity by season and species composition of vegetation, while problems of water supply were also assessed. More detailed investigation was conducted on the variation between pasture productivity and changes in vegetation composition with increasing distance from water points.

Measurement sites were selected around three settlements in the northern zone, and three in the central zone, each with their associated group of wells. Vegetation measurements, as described for the Yerbent regional transect, were taken at distances of 1, 5 and 10 km from the wells. In the southern foothill zone, two measurement sites were selected in the grazing areas of two farm associations. No vegetation measurements were taken around the settled canal zone, as the area of natural pasture is very limited.

Climate and soils along the regional transects

There is a slight precipitation gradient from north to south (Table 10.1). Precipitation is mainly in the form of rain occurring in winter and spring between December to May and reaches a peak in April. However, precipitation is variable from year to year and in some years can be only a few millimetres or none at all.

Winter temperatures are lower by several degrees Celsius in the northern part of the desert and by about one degree in the summer months. January is the coldest month with minimum temperatures of –28°C while summer temperatures can reach a maximum of 48°C.

The majority of both regional transect areas is covered by desert sandy soils which, although not saline, have a very low organic matter content of between 0.2–0.4 per cent. About 1 per cent of the total area is covered by *takyrs*, clay alluvial depressions that occur throughout the desert. *Takyrs* are very important features in the desert pastures as they collect rainwater used by livestock. There are also scattered areas of *solonchaks*, depressions composed of saline soils. These have no significance for land use. Soils in the canal and foothill zones vary from loams to sandy grey soil.

Topography varies from the northern area covered with sandy ridges (fixed dunes) with few *takyrs*, to the central zone of ridges with more and bigger *takyrs*. The canal zone is mainly flat, while the foothills in the southern area form rolling land with hills and rocky outcrops.

Vegetation and seasonal dynamics of forage availability

The rangelands within the desert transect zones are mainly classified as sandy desert, which represents the major type of rangelands in the country, with areas of

Table 10.1 Monthly and yearly precipitation for fifty years to 1998 (mm)

Climate stations and location	J	F	M	A	M	J	J	A	S	O	N	D	Annual total mm
Southern desert: Bahardok 38° 46' N 58° 30' E	16	19	21	27	13	3	1	1	1	11	9	15	141
Central desert: Yerbent 39° 18' N 58° 35' E	12	13	19	22	16	4	2	1	2	4	9	12	119
Northern desert: Darvaza 40° 10' N 58° 24' E	10	11	19	21	8	4	2	2	3	3	9	12	109

salt desert and, in the Yerbent transect, sand clay desert (Babaev 1994). Several main plant associations in the transect area were recorded in 1960 and again in 1985. These are:

- *Haloxylon persicum, Carex physodes* with *Ephedra stroboliacea*. The average total dry matter (DM) mass is 211 kg/ha while the available fodder portion is 88 kg/ha.
- *Haloxylon persicum, Carex physodes* with *Aristida pennata*. The average total DM mass is 239 kg/ha with an available fodder portion of 107 kg/ha.

A modified type was noted in 1985, due to grazing effects and removal of *H. persicum* by cutting:

- *Calligonum rubens, Salsola richteri, Carex physodes* with *Aristida pennata*. The average total DM mass is 187 kg/ha with an available fodder portion of 99 kg/ha.

Due to the low output of forage per hectare, stocking rates in the desert are low, varying from one sheep per six to fourteen hectares of pastures.

The sandy desert vegetation typically has a canopy of three layers. Upper layers of 1.20 to 1.50 m height are composed of large shrubs dominated by *Haloxylon persicum* and *Calligonum setosum* or *C. rubens*. The middle layer at 60 to 90 cm height is composed of dwarf-shrubs such as *Astragalus unifololatus* and *Convolvulus divaricatus* and tall grasses such as *Aristida pennata*. The lowest layer, with a maximum height of 25 cm, is formed of spring ephemeral annuals and perennials, with *Carex physodes* as the dominant species.

In the canal zone, most land is irrigated and cultivated, some with forage crops of lucerne and barley. In between and outside of the irrigated areas are small sandy hills used for grazing. The main pasture species are *Alhagi persarum* (camel thorn), *Phragmites comunis*, *Karelinia caspia*, *Atriplex* and halophytes on abandoned salinised land that has been irrigated. Camel thorn is a very valuable source of cut fodder for stall-feeding over winter.

Pastures of the foothill areas are the most productive, with an average of 90 to 480 kg DM/ha of available fodder. The main types of vegetation are *Poa bulbosa* and *Carex pachystilius* at a height of up to 30 cm. Tall grasses including *Astragalas spp.* and *Artemesia badhysi* are also found, together with *Salsola orientalis*.

Vegetation production and thus available forage varies by season, due to the seasonal precipitation pattern. As soils and topography are generally uniform throughout the sandy desert, locational differences in plant associations are mainly due to effects of grazing and cutting.

Particular types of plants contribute differentially to livestock forage in each season, as is shown in Figures 10.1, 10.2 and Table 10.2. One group of plants – ephemerals, comprising small herbs and grasses – contributes a proportionally higher percentage of its biomass to available forage as livestock can consume most of this group's biomass. The small sedge *Carex physodes* as well as other small herbs and grasses provide the greatest bulk of forage in all seasons. For example, in spring, 85 per cent of available forage can be from one species alone – *Carex physodes*. The

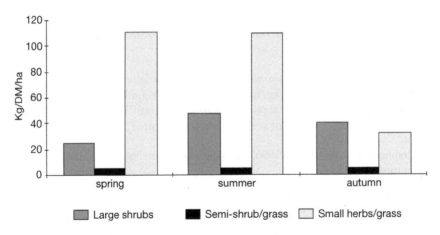

Figure 10.1 Total biomass of main types of desert vegetation, by season.

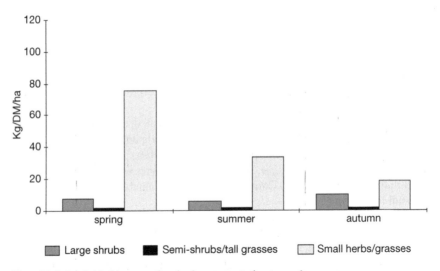

Figure 10.2 Available biomass of main desert vegetation types, by season.

late spring flush of *Carex* is very important for the productivity of desert-based livestock, as it becomes available after animals have over-wintered on poorer quality browse species and standing dead matter. The ephemeral spring desert flush provides livestock with a nutritional boost at the period of lambing and lactation.

However, the contribution of tall shrubs as browse rises in autumn and winter, as Table 10.2 indicates. Over 60 per cent of edible forage in autumn and winter is contributed by shrubs and dwarf-shrubs. In the cold dry season, the above-ground biomass of ephemerals such as *Carex* is dry and dead, while dominant woody shrubs of *Haloxylon* and *Calligonum* can still support browsing animals.

Table 10.2 Ratio of annual yield to available forage by season (per cent)

Dominant and major species	Spring 1st part	2nd part	Summer	Autumn	Winter
Haloxylon persicum	15/40	40/15	100/5	85/50	50/60
Calligonum setosum	0/0	100/80	80/80	20/50	10/50
Ephedra strobiliacca	50/50	80/15	100/5	85/50	50/60
Astragalus unifoliolatus	20/30	100/70	60/60	50/30	30/30
Convolvulus divaricatus	0/0	20/10	100/10	100/30	80/50
Aristida karelinii	50/10	75/10	100/0	100/5	95/10
Aristida pennata	50/10	75/10	100/5	90/25	75/50
Carex physodes	40/50	100/85	80/60	45/60	35/50
Heliotropium arguzoides	0/0	80/20	100/30	50/50	30/50
Bromus eremopurum	20/50	100/70	60/60	35/50	10/50
Isatis malcolmia	20/50	100/70	40/30	15/40	10/70

Note:
For each species the percentage of maximum annual biomass production present in each season is the left-hand figure. The percentage of the biomass that is potentially available in that season is the right-hand figure. For example, in the first part of spring, *Haloxylon persicum* reaches 15 per cent of its maximum annual biomass production, when 40 per cent is available to livestock.

Thus the period of maximum yield for some shrubs does not coincide with their maximum forage utility for livestock. In Table 10.2 it can be seen that *Haloxylon* reaches its maximum yield in summer, when only 5 per cent of that yield is available, whereas in winter when only half of the annual yield is present, 60 per cent is available to livestock. In contrast, the proportion of available forage for *Carex physodes* is relatively constant throughout the year, but is especially high in late spring. These two species have a particularly high nutritive value for livestock. In spring *Carex physodes* contains 126 g digestible protein per kilogram DM, while *Haloxylon* has 76 g/kg.

These seasonal dynamics formed the basis for seasonal livestock movement in the past. Woody shrubs were more prevalent in the northern desert compared to the southern desert, and animals could be taken there in winter to browse when ephemerals in the central desert were senescent. As is discussed below, recent changes in livestock management have meant that the winter northern migration is now curtailed.

Water resources for grazing areas

Water supplies in the pastures of Turkmenistan are of three main types; from the Kara Kum canal and its piped and drainage systems; from rain-fed clay depressions (*takyrs*) which may be improved by humans by building cisterns (*sardob*) and lastly from man-made wells. In total, about 72 per cent of the country's pastures are supplied with water of one type or another. However, only 44 per cent of pastures are supplied with year-round water supplies. Pastures with access only to salty wells may be utilised only in the cold season of the year, although the vegetation of these pastures could be used all year round. Since animals require less water in the cold season, they can be watered in winter from the salty water in these wells despite their high mineral content. Pastures supplied solely by rain-fed water sources can be used only in spring, as there is almost no rainfall in other seasons.

Access to water is the main factor limiting use of the pasture resources. An inventory of water points was carried out in both transects (Tables 10.3 and 10.4).

The availability of good quality water for livestock is unequally distributed within both transects. Yerbent association is divided into five farms, of which only two in the southern portion of the association's land are supplied by sweet water from the Kara Kum canal. The northern farms are supplied mainly by very salty and/or bitter wells. Unless tankers provide supplementary fresh water, livestock can only use these northern farms in the cold winter season when less drinking water is needed.

Table 10.3 Water resources for Yerbent transect, 1998

Type and number	Water condition	Construction conditions
Wells = 71 bores at 39 sites	Sweet water = 3 Slightly salty = 18 Strongly salty = 29 Highly salty and bitter = 21	Not functioning = 7. Few still have motorised pumps
Takyrs = 6	Rainwater	
Cisterns (*sardob*) = 2	Rainwater	Capacity of 10–12,000 litres
Piped water from Kara Kum canal, watering points = 24	Sweet water	

Table 10.4 Water resources for transect in Gok Tepe *etrap*, 1999

Ecological study zone	Type and number	Condition
Northern desert, 3 communities	25 wells Pipe line from canal to one community	All salty Not functioning
Central desert, 3 communities	12 wells Pipeline from canal to one community 3 cisterns (*sardob*)	All salty 2 not functioning Not functioning Almost dry in 1999 as little rain in spring
Canal zone	Kara Kum canal and its drainage channels	Infestation with parasites
Foothills, 2 farm associations	8 bored wells Mountain springs	Sweet water Sweet water

Some parts of the northern farms and pastures in both transects must be used year round, and are supplied by water tankers bringing canal water from the south of the association's land. Until the recent change of state farms into associations the state farms were responsible for transporting water to the northern pastures. However, the expense of transporting water now means that the associations are decreasingly able to carry out this service for their members.

Pasture use, water supply and degradation

National reforms in the latter part of the 1990s have resulted in the emergence of new systems of livestock management (Chapters 9 and 11). More livestock are privatised and control over animals is passed from the state farms to individual shepherding families. In tandem with the policy of privatising animal ownership is the reduction in state financial support to the former state farms, which are now expected to become self-supporting farmers' associations.

These two changes have several immediate effects on grazing patterns and water provision in the desert pastures. First, since the state farms were converted into associations, government financial support has been reduced. Privatised shepherds no longer receive state assistance for the seasonal movement of animals in their care from one pasture to another. If animals are to be moved, the shepherd must find the means and pay the costs.

Second, the associations now lack financial resources to transport sweet water from the canal or pipeline by tanker to distant pastures that are only supplied by salty or bitter wells, or have no water at all. Consequently, only the central desert area remains regularly and inexpensively supplied with sweet or only slightly salty water. Shepherds must pay for transporting water if they want to graze their animals in remoter pastures.

Third, the government body responsible for maintaining wells, pumps and pipelines is less effective than in former times. As noted in Tables 10.3 and 10.4 some wells and pipelines are no longer functioning. With the change from state farm control to more privatised water management, some pumps have been removed from the desert wells and taken away by officials to other areas. Privatised shepherds who have enough financial resources are now buying their own pumps as well as paying for water to be transported to remoter pastures. This is occurring particularly in the northern desert, where, though settlement is very sparse, families have more livestock resources (see Chapter 11). But the majority of shepherd families in the central zone have fewer livestock assets and cannot afford to supplement water supplies.

These new limitations on grazing and water management are leading to increasing concentrations of people and livestock around better-quality and cheaper water sources in the central desert. The radius around settlements with wells is being grazed more heavily, while large areas of pastures in the north are underused where ground water is more mineralised and fresh water costs more to transport.

The greater concentration of livestock in the central zone, compared to the north, is directly related to the presence of year-round livestock drinking water. In Yerbent farm association, some of this water is provided free, piped from the Kara Kum canal. Even when fresh water for animals has to be transported by tanker, it is cheaper for private shepherds to move this water a shorter distance. Thus, in both Yerbent and Gok Tepe desert areas, livestock are now generally kept within a day's walk of wells, a maximum of 12 km from villages. Only in winter can animals be moved further away, up to 20 km from water points, as they require less water in the cool season and can be watered from tankers.

Heavier grazing around water points in the central zones results from the fact that most animals are now privately owned and the farm association has no obligation to provide water for these animals (see Chapters 9 and 11). An increasing number of privately owned animals are not moved to the northern pastures for winter, as their owners cannot afford the costs of supplying water to them there. Instead, more animals are being kept for most of the year around wells in the central desert.

Inevitably as grazing pressure builds around wells in the central desert, the vegetation is being affected. Measuring changes in vegetation is a long-term process. It is not possible on the basis of a one- or two-year study to assess changes in the vegetation leading to land degradation at a particular site. Degradation may be defined as one or more of the following: as a loss of desirable pasture species, a reduction in plant cover, a reduction of species diversity or a loss of top soil.

The results of this study in two transect areas of the Kara Kum desert should be considered as a baseline for future monitoring of long-term vegetation and land degradation trends. Nevertheless, it is possible from the available data to reach some preliminary conclusions on changes in plant species by comparing results from more heavily and less heavily grazed locations. These results suggest that longer-term processes of degradation are already underway in certain areas, due to grazing pressure.

Degradation due to overgrazing can be avoided by the appropriate balance between stock numbers and pasture, and by appropriate seasonal movement of animals between pastures. By moving, or rotating, animals onto different grazing areas, over-concentration in one area long enough to cause damage to the vegetation may be avoided. As discussed previously, the seasonal and locational variation in forage plant productivity also provides a rationale for moving animals to different pastures. This well-recognised management strategy was employed until recently by the state livestock farms, and before them, by the Turkmen nomads.

Under the current farm association administrations, these seasonal movements are being curtailed. Yerbent state farm and its new form of association had a plan for the seasonal movement of livestock to different pastures, but this plan cannot realistically be implemented due to the uneven distribution of good water resources. Likewise, in Gok Tepe district, among the eight associations keeping livestock in the desert, there is no longer any attempt by the administrative authorities to control where animals are moved in each season. Seasonal movements to different pastures are now determined by individual shepherds, according to where well water is least mineralised.

As more livestock can be watered year-round in the central zones and irrigation is available for crop production, the concentrations of people are also higher. This has led to cutting of the dominant woody shrub *Haloxylon persicum* for household fuel. Gas for domestic fuel is not routinely supplied to the desert households. The proximity of the central desert to the more densely settled areas to the south along the Kara Kum canal also means that people from the canal zones come into this part of the desert to cut *Haloxylon* and other woody shrubs for fuel wood.

As can be seen in Figure 10.3, *Haloxylon*, the dominant shrub in the sandy desert region, was not found at least within 10 km of settlements in the central desert zone. From Figure 10.4 it can be seen that this shrub remains in the less-settled northern desert (although at a reduced biomass within 1 km of settlements), where the biomass of woody shrubs overall is about three times higher.

The loss of woody shrubs has two main negative impacts on land degradation in the desert environment. First is that autumn and winter forage availability is reduced, as shrubs form the main forage resource in these seasons. In winter, 50 per cent of the total biomass production of *Haloxylon* and *Calligonum* is available to livestock (see Table 10.2). The second problem resulting from loss of shrubs is greater instability of the sand dunes around settlements. The desert shrubs are very deep-rooted and have an important function in stabilising sand dunes. Once these

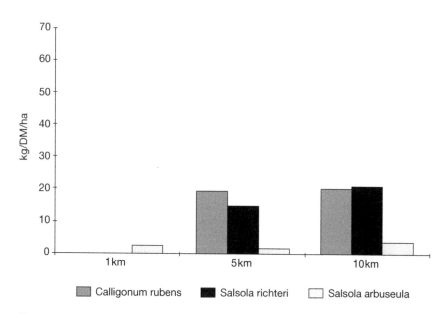

Figure 10.3 Total biomass of shrubs at different distances from water, central desert, summer 1999.

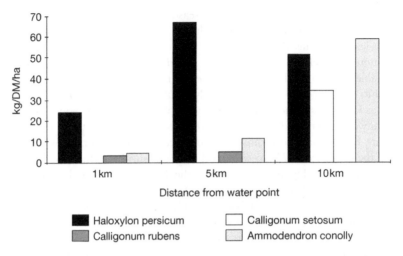

Figure 10.4 Total biomass of shrubs at different distances from water, northern desert, summer 1999.

shrubs are removed, sand builds up into moving dunes, which can seriously encroach into desert settlements.

Heavier grazing pressure in the central desert zone as a result of changes in livestock management is also creating other signs of undesirable changes in vegetation, with desirable forage species being replaced by less desirable, nutritionally inferior ones. If a comparison is made between northern and central zones in terms of vegetation composition radiating out from water points, certain differences in vegetation composition are clear.

Earlier in this chapter, the importance of *Carex physodes* as the main constituent of spring and summer forage was discussed. This species is retreating from the accessible pastures of the central desert, as Figure 10.5 and Table 10.5. show. Here it is important to recall that precipitation is on average only slightly lower in the north, while soil and topography is uniform between the north and central desert. The vegetation associations should therefore be comparable in the north and central desert, and changes noted are therefore likely to be due to grazing.

The main spring forage plant, *Carex physodes*, is completely absent close to water points in the central desert (Figure 10. 5). In the northern desert, this plant still makes up nearly 30 per cent of spring forage even on pastures 1 km from water points. Further away from water, *Carex* in the northern pastures comprises almost 80 per cent of spring forage while only reaching a maximum of 47 per cent in the central desert pastures. These results suggest that central desert pastures have lost some of their best quality and most important forage species. Northern pastures are at risk also, as *Carex* is less commonly found closer to water points there. How long this process has been occurring could not be assessed in this study, but should be investigated in future work.

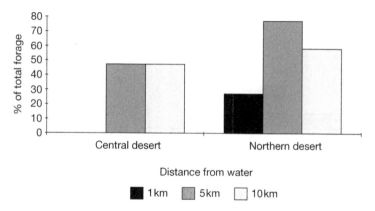

Figure 10.5 Change in percentage of *Carex physodes* in available forage between northern and central deserts, by distance from water, spring 1999.

Table 10. 5 Available forage biomass (kg DM/ha) by distance from water, spring 1999

Plant species or group	1 km from well	5 km from well	10 km from well
Central desert			
Calligonum rubens	0	14.7	17.5
Salsola arbuseula	0.1	0	0.2
Salsola richteri	0	0.5	0.6
Aristida karelinii	0.9	0	0
Astragalus longipetiolatus	0	2.0	0
Peganum harmala (in summer)	0	0	0
Other dwarf-shrubs and tall grasses	0	0.1	4.4
Aristida pennata	0	0.6	1.0
Carex physodes	0	24.0	29.0
Grasses	0.2	35.0	25.2
Total (kg/ha)	**1.2**	**76.9**	**77.9**
Total fodder units	**0.3**	**64.1**	**54.3**
Ground cover	**5%**	**30%**	**30%**
Northern desert			
Haloxylon persicum	1.4	4.1	2.9
Calligonum rubens	2.3	3.5	25.9
Ammodendron conolly	0.1	0.3	1.5
Shrubs in total	*3.8*	*7.9*	*30.3*
Astragalus longipetiolatus	0.4	0.8	34.6
Aristida pennata	0.8	0.4	4.9
Carex physodes	5.2	48.5	67.0
Grasses	8.5	4.3	8.0
Total (kg/ha)	**18.7**	**61.9**	**144.9**
Total fodder units	**15.2**	**53.7**	**120.2**
Ground cover	**20%**	**30%**	**40%**

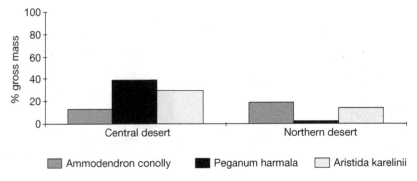

Figure 10.6 Percentage of total biomass of unpalatable species in northern and central deserts, at 1 km from water points, summer 1999.

Loss of desirable forage species due to overgrazing is often accompanied by an increase of species that are of little or no nutritional value and are unpalatable to livestock. The encroachment of unpalatable species is shown in Figure 10.6. These species are mainly *Aristida karelinii, Peganum harmale* and *Ammodendron conolly*. It was found that, with one exception, these unpalatable species occurred only within immediate proximity of the water points within the settlements, where grazing pressure is predictably the greatest. The exception is *Ammodendron conolly*, which was present at all three northern desert sites, both near to and further from water points. However, the other two unpalatable species were absent at two of the three northern desert sites.

At the three central desert sites, unpalatable species dominated within one kilometre of the water points, but were rarely present at five and ten kilometres distance from water.

These patterns suggest that heavy grazing pressure in the central desert has caused more unpalatable species to appear near to settlements around water points. But further away from water, unpalatable species are not yet common either in the central or northern desert. Since animals are usually taken for grazing further than one kilometre from settlements, the appearance of unpalatable species close to water points probably does not present a major problem for livestock management. There is nevertheless a risk, from the observed pattern, that should grazing pressure increase, unpalatable species may spread further away from water points as more palatable species are gradually overgrazed in a widening circumference around water points.

Finally, we consider the overall effect on forage availability caused by grazing pressure. It has been suggested that grazing pressure is leading to the disappearance of more nutritious forage plants, their replacement with lower-quality forage species and a lower available plant biomass per area. The effect of these processes can be measured by looking at the total fodder values of the available forage plants (Figure 10.7). Fodder values are explained in Appendix I, page 257.

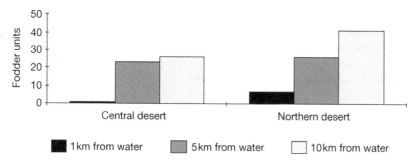

Figure 10.7 Pasture fodder units at different distances from water points in northern and central deserts, summer 1999.

Not surprisingly, the fodder values shown in Figure 10.7 reflect the relative lack of more valuable forage species in the central desert, particularly in the pastures closest to water points. An interesting result is that in the northern desert, fodder values continue to rise with distance from water, whereas pasture fodder values level off at the 5 km mark in the central desert.

One explanation could be that shepherds have settled the central desert areas and grazed their animals there for a much longer period than in the northern desert. This would result in a general lowering of plant productivity due to long-term grazing pressure in almost all areas. A second explanation could be that the population of camels is higher in the northern desert communities (see Chapter 11). With their capacity for foraging much further distances from water sources than smallstock, camels are able to disperse the grazing pressure to a much wider area away from water sources. In this case, it is possible that forage availability in the northern desert continues to rise beyond 10 km from water sources. Future studies will need to investigate these two possibilities as well as other explanations for the pattern of vegetation in response to grazing.

Conclusions

Turkmen nomadic pastoralists have used the desert pastures for many centuries. The vegetation has therefore long been modified by the impact of livestock grazing. Starting before the middle of the twentieth century, the livestock management systems have been changed by state policies that have had, and will continue to have, further impacts on the vegetation.

During the period of collective state farms, boundaries were imposed on the seasonal movement of livestock and state farm officials regulated these movements. The traditional water harvesting technology used by Turkmen pastoralists was enhanced and new higher capacity water sources provided. Other inputs, in the form of supplementary winter feed, transportation and veterinary health services, allowed livestock populations to increase. Nomads were encouraged to settle around new service centres in the desert, while livestock, then under state ownership, were rotated in each season to different pastures. Concentrations of people and animals built up around wells and piped water supplies from the Kara Kum canal.

However, water supplies remained a constraint to pasture use in the northern part of the desert, where the ground water is heavily mineralised. As sheep can only drink small amounts of this water, the period in which they could graze the northern pastures was limited by the scarcity of harvested rainwater supplies. By transporting supplementary fresh water in tankers, the state farms were able to keep more flocks of sheep as well as camels in the northern desert. In autumn and winter, when livestock required less drinking water, more animals were also moved long distances from the southern and central desert to the northern pastures where they could browse on shrubs. This seasonal migration system eased the grazing pressure on more heavily stocked central and southern pastures.

Over decades of state farm grazing and water management, the composition and density of vegetation has again been altered in all parts of the desert accessible to livestock. More animals can be kept around the larger settlements with good supplies of sweet water in the central desert. The grazing pressure from these animals and the cutting of shrubs for fuel has left a circle of nearly denuded land in the vicinity of these villages, where the main plants that now occur are unpalatable to livestock. Further away from the central desert villages, but still within a day's walk for sheep, the quantity of forage plants has also been reduced compared to the northern pastures.

In the northern pastures, stocking levels have not been able to increase to the same degree, due to the shortage of sweet water for livestock. Although some negative effects of grazing pressure are found in the immediate vicinity of northern desert settlements, the density of shrubs is higher and overall these pastures contain more forage for livestock.

The reforms in state farm organisation introduced in the middle of the 1990s are leading to more changes in grazing and water management, as shepherds adjust to privatised livestock ownership and state support services are withdrawn. These changes can be expected to further alter vegetation composition and density. While it is too soon to measure the impact of these changes, it is possible to detect certain trends. Additional and longer-term measurements will be needed to determine how the reforms will ultimately affect the natural resources of the desert.

At present, the main trend is that livestock grazing pressure is building up around settlements in the central desert, as animals are moved less than before. The costs of moving animals to more distant pastures, and providing water for the animals once there, must now be borne by the newly privatised shepherds. As a consequence, shepherds are relying more on cheap or free sweet water supplies that are found in the central rather than the northern desert.

Under the new management regimes the pattern that developed during the state farm period, whereby central pastures were more heavily used than northern pastures, is therefore being reinforced. Higher grazing and human pressure has already led to a loss of useful vegetation resources in the central desert, and this trend is likely to continue as a result of the privatisation process.

To lessen the risk of further land degradation, it will be necessary to rehabilitate existing non-operating water sources, and to increase the area of pasture with access to water by providing new small-capacity water points, such as the traditional water cistern (*sardob*) which collects rainwater. These measures would allow grazing pressure to be dispersed further away from the settlements where animals are now concentrated. This would reduce the grazing pressure around settlements. Whether the vegetation in these areas would recover naturally or whether other forms of intervention would be required remains to be determined, but the risk of future degradation would be reduced.

11

NEW PATTERNS OF LIVESTOCK PRODUCTION

Ovlyakuli Hodjakov and Iain A. Wright

Introduction

Though gradual, the process of reform in Turkmenistan has already had a number of significant effects on livestock production. Two principal effects can be discerned. There has been a shift of livestock management away from control by state farms to individual households, which has meant adjustments in the way livestock are managed and in their output. Second, state support to livestock production has been sharply curtailed, reducing access to critical infrastructure and inputs. This has threatened the economic viability of many livestock production units and increased their vulnerability to sudden climatic shocks as well as to market forces beyond their control. This chapter compares the impacts of reform in two contrasting institutional settings, and shows how national policies have altered local-level management of livestock.

Agricultural policy

The Turkmenistan government has a policy of food security which is to be achieved through self-sufficiency. For example, annual wheat production in 1997 was 1.24 million tonnes, but the goal is to produce 2.0 million tonnes to ensure adequate supplies of flour for bread. Each family is supposed to receive a monthly allocation of flour from the state, although many families do not. The policy of food self-sufficiency affects other sectors, as land is being switched from cotton to wheat and little wheat is available for animal feed, although arable by-products, such as wheat bran, straw and stubble, are available as fodder. There are plans to extend the use of desert areas by providing watering points in areas which currently have no water. There is also a national policy to increase sheep numbers (in the state sector) by 5 per cent each year until 2002.

Under these plans, grain area has more than tripled since 1990, and the area under wheat has increased ninefold. This fact alone has also had an adverse effect on livestock production since it has happened at the expense of fodder crops for animals which were reduced from 337,000 ha in 1990 to less than 245,000 ha in 1996.

The mechanisms for the delivery of agricultural policy are difficult to ascertain. Much of the policy is set by presidential decree with the means of implementation not always being clear. Ministries are in a continual state of flux, with new ministers appointed and amalgamation of Ministries. Organisations such as Turkmen Mallory (literally Turkmen Livestock) have been established to support the activities of the new associations formed from state farms, in different areas of activity (see Chapter 9). There are equivalent organisations for cereals, vegetables, etc. Some research institutes have been transferred to the control of these organisations, e.g. the Institute of Livestock and Veterinary Husbandry was transferred to Turkmen Mallory in 1998, and transferred again in 1999 to the Ministry of Agriculture.

Details of the re-organisation of state farms into associations are provided in Chapter 9. Briefly, in 1995 the state farms were abolished to be replaced by associations to which the former state farm employees could belong. Assets, but not land, were transferred to the associations. However, in 1998 the President announced that 51 per cent of the association shares were to be transferred back to the state. The government still buys produce from these associations (see Chapter 13). The associations can lease livestock (and crop land) to individual families. Leaseholders have target production levels and must return a proportion of the output back to the association but retain the remainder.

Pasture land

Land in Turkmenistan is mainly pasture – 78 per cent of land, equal to 38.3 million hectares of which 26 million hectares are classified as desert (Babaev *et al.* 1991). A further 1.8 million hectares are irrigated, used for growing mainly wheat and cotton. It has been estimated that a total of 16 to 18 million hectares could be used for growing crops if they could be irrigated. The deserts are characterised by low rainfall, high summer temperatures and low plant productivity. The vegetation has been classified according to plant community and mapped. Generally, three major groups of desert plant communities are recognised. Moving from the driest and hottest climatic zones, they are: (a) bushes (b) dwarf-bushes and long grasses, and (c) short grasses.

Generally, plant productivity in the desert pastures is low, from 150 to 200 kg dry matter (DM)/ha/year, although in some years it can fall below 100 kg DM/ha/year. Annual production is limited mainly by rainfall, although grazing intensity certainly has a local effect. There are 2.3 million hectares of mountain pastures. The pastures in the foothills are grazed, but most mountain pastures are not used by livestock, although some private shepherds now graze the lower hill pastures.

Water availability

Only 60 per cent of the desert pastures have wells and can therefore be utilised by livestock. The proportion of pastures with wells is less in the north of the country. For example, only 30 per cent of Dashauz region is supplied with wells. Even if water is available in some areas it is sometimes too salty for consumption by livestock. Some areas are supplied with water by pipeline from the Kara Kum canal. Further information on water availability is given in Chapter 10.

In summer, sheep need two to eight litres of water per day, with an average of five to six litres depending on their size, and need water every day or every second day. In winter, water consumption is less as temperatures are much lower, and sheep do not need to be watered every day. Because water consumption is lower in winter, livestock, and sheep in particular, can consume water with a higher salt content than in summer.

Livestock numbers

National statistics on livestock numbers are collected by the government's National Institute of Statistics. However, it was admitted by senior staff that efforts are concentrated mainly on government sector statistics, because information on the private sector of the economy was difficult to obtain. Data on livestock numbers are collected from a sample of shepherds. The shepherds in the sample complete forms and the numbers are then multiplied up to give a figure for each region and then nationally. The numbers of livestock in each sector is shown in Table 11.1.

The true numbers in the private sector are likely to be higher than the figures in Table 11.1 because shepherds are reluctant to divulge numbers of livestock that they own privately. There is a traditional suspicion of government organisations, dating from the Soviet era when there were upper limits to the number of animals which were permitted to be owned privately. Also, shepherds fear that divulging their true numbers of sheep might lead to their having to pay taxes. When national livestock experts were asked how many sheep and goats were in Turkmenistan, estimates ranged from eight to ten million.

An additional issue is that confusion surrounds the legal ownership of the associations' livestock. Many people report that the stock belongs to the state. Indeed, for the purposes of the livestock census these animals are regarded as state livestock. Others indicate that the livestock are owned by Turkmen Mallory, while

Table 11.1 Livestock numbers (thousands) in Turkmenistan at 1 January 1998

Sector	Sheep and goats	Cattle	Camels
State	3,131.1	216.1	50.8
Private	3,345.0	998.1	52.1
Total	6,476.1	1,214.2	102.9

Source: Turkmenistan National Institute of Statistics.

yet others claim that the animals belong to the associations which have replaced the state collective farms.

Sheep breeds

Karakul

Karakul (Astrakhan) sheep are kept for the production of pelts and for meat. Karakul sheep are indigenous to Central Asia and highly adapted to the desert environment. Karakul pelts were a major product from Turkmenistan's livestock sector in Soviet times, when more than one million pelts were exported annually. By 1998 this figure had dropped to less than a quarter million, due to lower demand from the Russian market, while quality control of production and processing has also declined. The majority of sheep in Turkmenistan are Karakul, amounting to some four million (Bruzon 1998). However, in the two study areas of the Kara Kum desert, Karakul are not as commonly kept as the Saraja sheep, described below.

Karakul pelts are produced when the lambs are slaughtered at two to three days old. If slaughter is delayed until ten days, the hair has started to grow and the quality of the pelt is greatly reduced. Foetuses are sometimes aborted at eighteen to twenty days before natural birth to produce pelts of higher quality. Good quality Karakul pelts should have small, tight, hard waves, lying parallel to one another. The name Karakul means 'black lake', referring to the wavy effect. The hair should be 7–8 mm long. Longer hair (e.g. 12–15 mm) results in poorer quality pelts. It is claimed that if Karakul sheep are fed too well in pregnancy, the lambs have thicker skin and longer hair and this reduces the pelt quality. Animal breeders have concentrated on improving the quality of pelts. Heritabilty (h^2) of quality is quoted as being 0.2–0.4, although it is not clear how this is measured objectively. There are four colours of Karakul produced in Turkmenistan, described in Table 11.2.

Mean live weights of adult Karakul sheep are 42–45 kg for ewes and 60–70 kg for rams. Lambs are 28–32 kg at three months old. In the desert areas the sheep are 8–10 per cent lighter. Typically lambing percentage is about 90 per cent but, if well fed, they can produce up to 120 per cent. However twins are regarded as undesirable because of their smaller size and poorer quality skins. If mixed-sex

Table 11.2 Types and prices of Karakul pelts

Colour	Description	Percentage of Turkmenistan flock	Price per pelt US$ *
Black	All hairs black	75–80	7–8
Grey	Mixed black and white hairs	12–13	10–12
Brown	Brown hair	5–6	n/a
Sur	Dark hairs with light tips	5–6	15 (for good quality)

Note
* Prices quoted by Karakul breeder for 1998.

twins are born, then the male lamb is killed and the female reared but if the twins are both female, then one is usually fostered on to a ewe whose male lamb has been killed for its pelt.

Karakul sheep normally lamb for the first time at two years old, although if well fed, 70 per cent can lamb at one year old. However, lambs born to yearling ewes are usually slaughtered for pelts to avoid the drain of lactation on these young sheep. Adult Karakul sheep are shorn, but the wool is coarser than from Saraja sheep.

Saraja

The Turkmen consider that wool from the Saraja sheep makes the best carpets in Central Asia. Saraja sheep are a white, triple-coated breed, with three distinct types of hair (Table 11.3).

There should be no more than 1–2 per cent coloured fibres in a Saraja fleece and the fine fibres must be at least 5.5 cm in length for processing. The Saraja gives 15–20 per cent more meat that the Karakul, because it is heavier, with a mature live weight of approximately 50 kg.

Saraja sheep are considered to be less hardy than the Karakul, and less well adapted to desert conditions. They can only walk up to 15 km per day whereas it is claimed that Karakul can walk 20 km per day. Saraja sheep lose more live weight in winter and mortality in winter is higher. The lambing percentage of Saraja sheep is usually 85–88 per cent, although under good nutritional conditions they will produce a lambing percentage of 110–115 per cent. The Institute of Livestock and Veterinary Husbandry is aiming for higher lambing percentages with a target of 115 per cent. However, this level of productivity can only be supported in the foothills of the mountains where there is enough fodder to support a proportion of ewes with twin lambs.

Methods of study

Field work was carried out in the two study transects of the research project: first, the large farm association of Yerbent, and second, Gok Tepe *etrap* (administrative district). Further information on these areas is given in Chapters 9 and 10.

In spring 1998 interviews were conducted with fifty livestock-rearing households in three settlements of Yerbent association. The settlements were selected along the north–south axis of the study transect, located at increasing distances from

Table 11.3 Fibre types in Saraja fleece

Fibre type	Diameter (mm)	Percentage
Down	20	80
Medium	38–40	15–18
Rough	57–8	4–5

Ashgabat at 110 km, 160 km and 240 km from the city. The latter settlement is the most remote, being 40 km away from the main highway, a five-hour drive across the sand dunes from the main road. In addition, livestock production data were obtained from the professional staff of Yerbent association.

In summer 1999, interviews were carried out in Gok Tepe district with forty-five livestock-rearing households. The sample was selected along the north–south axis of the study transect, in four agro-ecological zones consisting of foothills, canal zone, central desert and northern desert, at increasing distances from the population centres of the canal zone and Ashgabat. Data on livestock were also obtained from the district professional staff in Gok Tepe town.

Livestock production in Yerbent

Resources of Yerbent farm association

The total area of the association is 863,000 ha of which 98 per cent is pasture land, with about 900 ha of irrigated land. There are 520 workers, with 472 working in agriculture, of which 407 work in the livestock sector. The association has five farms, one specialising in camels (Charan), and four specialising in sheep (Mametyar, Yerbent, Bori and Sulfur-plant settlements) although camels are also kept at Bori. In 1998 the farm grew 120 ha of wheat, which produced 23,000 kg of grain, giving an average yield of only 191 kg/ha. Some of the grain was given to the state in return for a combine harvester. Of the 120 ha of wheat, straw was produced from 65 ha (70 t), while straw was not harvested from the remaining 55 ha which was however grazed by sheep. In 1998 300 ha of alfalfa were grown of which 200 ha had been sown in 1997 and 100 ha in 1998.

Details of the pasture types in Yerbent association land are given in Chapter 10.

Livestock numbers

The association's livestock is shown in Table 11.4. All the sheep owned by the association are Karakul. Yerbent is the only association in the region to have Karakul sheep, which were introduced in the mid-1980s as an experiment. The camels are all of the single-humped Arvana breed, kept for milk and meat.

Table 11.4 Livestock numbers in Yerbent association at 1 January 1998

Class of stock	Sheep and goats*	Camels
Breeding females	23,895	954
Other stock	25,917	1,323
Total	49,812	2,277

Note
* Officially, sheep and goats are classified together. As far as could be ascertained there were no goats owned by Yerbent association.

The members of the association also own private livestock. In 1998 there were approximately 2,300 private camels and 20,000 private sheep. These camels are kept primarily for milk production, the milk being used to make fermented milk products such as *chal* for family consumption. All private sheep are of the Saraja breed. Private goats kept for meat were also seen on the association but no estimate of their numbers was available. Also some private cattle are kept, but again no estimate of numbers was available.

In the past the state farm had over 70,000 sheep, but this was reduced to 50,000 as a consequence of a previous five-year plan to increase the number of Karakul pelts produced. The increase in production was achieved at the expense of total numbers, by slaughtering many lambs that would normally have been kept as breeding replacements. In keeping with present national policy decreed by the President, the association was then required to increase the number of sheep by 5 per cent each year until 2002. There are no production targets to be met, only an increase in the total number of sheep held by the association. The increase in numbers was achieved by keeping older sheep and by slaughtering few lambs for Karakul pelts. Thus the income to the association was severely affected, as it could not sell as many pelts as in the past. Interestingly, a former association director would have preferred to reduce the total number of sheep to 40,000 and dispose of old, unproductive sheep and younger, poor quality sheep. This would have allowed the association to concentrate on improving the quality of the stock.

Sheep production

In 1998, 23,900 ewes produced 22,000 lambs, i.e. a lambing percentage of 92 per cent. The lambing percentage in 1997 was 90.8 per cent. Note that these figures are based on the number of lambs born alive as a percentage of the ewes present at the beginning of the year, not as a percentage of those mated, which is the more common way of expressing lambing percentages in Europe. Only 5,760 lambs were slaughtered for Karakul pelts, and these were mainly weak and sick lambs, and 7,950 female and 8,280 male lambs were kept for rearing, both as replacement stock and for rearing for meat. Only 4,700 of these male lambs were needed as replacement breeding rams.

Sheep management

The sheep owned by the association are divided into about 80 flocks of 700 head, with about 20 flocks on each of the four sheep farms. Each farm has its separate spring–summer and autumn–winter pastures. Usually summer pastures are near wells because the sheep have to be watered every day, while winter pastures can be further from wells since sheep need water every second or even third day in winter (further details are provided in Chapter 9).

At the beginning of February the sheep are moved from the winter pastures to the summer pastures, although in some farms there may be special lambing pastures. For example, at Yerbent settlement, the lambing pastures are 6–10 km from the settlement. After lambing the sheep are transferred to the summer pastures. Growth of annual plants and shrubs begins in March, coinciding with the start of lambing and so providing an increased level of nutrition to support milk production. Lambing takes place from 10–15 March until April. Lambs are 3.5 to 5.2 kg at birth, with the mean birth weight of males and females being 4.3 and 3.9 kg respectively. Lambs are weaned in July, and flocks are re-organised. Lambs receive the best pastures, with many of them grazing wheat stubble.

After shearing in August–September the sheep are moved to their autumn–winter pastures. In winter the grazing sheep are supplemented with hay (made from both natural pasture and sown pasture such as alfalfa), straw and concentrate supplement. Each farm is supplied with enough hay for twelve months and with concentrate for seven to ten days. If snow cover is deep, sheep are moved to simple shelters and fed.

In summer, some of the sheep are grazed on land controlled by another association in neighbouring Gok Tepe district.

Each flock is centred on a well or group of wells. At Yerbent settlement, the wells are from 10 to 40 km from the settlement. Each flock of sheep is looked after by a brigade of a few shepherds who take it in turns to accompany the sheep. The length of time spent with the sheep is agreed by each brigade, but is usually one to two weeks. The shepherds use motor bikes to travel to and from the settlement to the flock, staying in small shelters for the duration of their shift.

Camel management

Yerbent association had 2,277 camels (at 1 January 1998). The Arvana breed are relatively small single-humped camels. One of the farms (Charan) is a specialist camel farm, while Bori in the north also has 500 camels. No camels are leased by the association to individuals because, according to the director, this would be too difficult to control, because it would be impossible to identify the leased camels. However most people own private camels.

Camels calve in March after a thirteen-month gestation. Normally they calve every two years and are milked for a year. At the specialist camel farm (Charan) the camels roam freely in summer, returning to the well for water themselves, but in winter they are herded.

At Bori settlement the 500 camels belonging to the association are managed by a head camel herder assisted by twelve herders. These camels are not milked. In summer they roam the desert and are gathered by motorbike in October. Some camels wander off the association's land and are brought back by other shepherds (the camels are all branded) for payment in October. In winter, camels are hobbled and are kept within 15 km of the settlement.

Young camels are weaned naturally at about one year of age. There is no planned marketing strategy. Depending on how many camels have been born, older camels are sold for meat and so the number sold each year is very variable. In some years 100 camels are sold, while in others only ten or twenty are sold.

In 1996, no young camels were born. The she-camels aborted several months before the expected date of parturition and also had skin lesions. The association's veterinarian could not diagnose the condition, but the level of his qualifications is not known. Many veterinarians working for associations have little, if any, formal veterinary qualifications. However, since that year the problem has not recurred.

The camels are shorn and the wool sent to the association's central administration, although no camel wool was sold in 1998 because no buyers could be found.

Provision of fodder

The association makes hay from natural pastures and lucerne. Coarse fodder is hay made from a number of stemmy herbaceous plants and shrubs, e.g. camel thorn (*Alhagi persarum*), while hay is made from lucerne, natural grasses and herbs. In addition, straw is available from the cereal crops. The amounts of fodder fed in 1997 were 2062 t coarse fodder, 1629 t hay, 30 t wheat straw, 293 t of mixed concentrate and 5 t grain.

Animal health

The main infectious disease is sheep pox, against which sheep are vaccinated. Some cases of pox have occurred in camels in the last three years. Affected animals are isolated and treated with antibiotics.

Brucellosis tests are conducted and, if reactors are found, the animals are isolated and slaughtered for meat. Fecal analysis for internal parasites is conducted and, if necessary, anthelmintics are used, although this is only necessary in the irrigated area near the Kara Kum canal. Pneumonia can be a problem in sheep which are watered from wells with acrid water in summer. When this occurs affected animals are isolated and should be treated with antibiotics, although in the past few years financial difficulties have made the purchase of antibiotics difficult. Sheep mortality was about 5 per cent in 1997.

Private animals

In Yerbent association all privately owned sheep are of the Saraja breed. The reason for this, according to the director, is that it avoids confusion with the association's Karakul sheep.

Information was collected on management of private stock from interviews with fifty households in three settlements of Yerbent association. Table 11.5 gives details of the numbers of private sheep owned by these households. Many people also had between one and six camels.

The further north into the desert and the further from Ashghabat, the larger were the private flocks. In particular, at Bori which is 240 km from Ashghabat and 40 km from the nearest asphalt road, private flocks were large. In this settlement,

Table 11.5 Sizes of private flocks in three farms in Yerbent association

Settlement	Distance from Ashgabat	No. of h/holds surveyed	Distribution of families according to number of sheep owned (and percentage)				
			0	1–25	25–50	50–100	100–200
Mametyar	110 km	17	4 (23%)	9 (53%)	2 (12%)	2 (12%)	–
Yerbent	160 km	22	–	14 (64%)	3 (14%)	5 (23%)	–
Bori	240 km	11	–	1 (9%)	3 (27%)	5 (46%)	2 (18%)
Total		**50**	**4 (8%)**	**24 (48%)**	**8 (16%)**	**12 (24%)**	**2 (4%)**

nearly 65 per cent of families had 50–200 private sheep and goats, while in Mametyar 24 per cent of families do not have any livestock and 23 per cent have only between one and twenty-five sheep and goats. These figures must be treated with caution in that they may underestimate the numbers of private livestock because people are wary about divulging the number of private livestock they own, as previously noted. Nevertheless the general trends should be valid.

Table 11.6 shows estimates by respondents of the amount of feed given to private sheep and goats.

The shepherds who fed concentrate bought it either from the association or from commercial sources. The mean amount of natural hay fed per animal was 36.5 kg while the mean amount of concentrate fed was 24 kg. Generally those with smaller flocks fed higher levels of both hay and concentrates. There could be several reasons for this pattern. Households with smaller flocks may be trying to increase

Table 11.6 Amounts of feed fed to private sheep and goats in winter (kg/head) by households in sample, Yerbent

Settlement	Fodder type	Number of sheep per household			
	kg/head	1–25	25–50	50–100	100–200
Mametyar	Concentrate	19	22	22	–
	Natural hay	57	44	25	
Yerbent	Concentrate	37	–	19	19
	Natural hay	75	–	57	31
Bori	Concentrate	33	42	32	18
	Natural hay		0	0	0

their numbers, while those with larger flocks also have the resources to move animals to more distant pastures, thus reducing the need for supplementary feed. In the furthest north settlement of Bori, none of the private flocks received hay, because of the availability of shrubs and dwarf shrubs in this more remote desert location which can be grazed in winter and in which there is less collection of firewood, as noted in Chapter 10.

A comparison of livestock feeding regimes by Yerbent association versus private owners reveals that privately owned animals receive on average three times as much concentrated fodder per head (24 kg/head) than animals belonging to the association (6 kg/head). The reverse is found for feeding of natural hay – private animals receive 37 kg/head compared to 45 kg/head for association animals. Since concentrates are more expensive and nutritious than natural hay, this pattern suggests that private owners are more able or willing to expend financial resources to provide better nutrition for their animals. This issue is further discussed in Chapter 12.

Livestock production in Gok Tepe

Resources of Gok Tepe etrap (district)

Gok Tepe district has a total area of 572,600 ha, of which 81 per cent (462,500 ha) is pasture and 3 per cent irrigated. The remainder is non-agricultural land in the mountains. Overall there is 4.8 ha of pasture land per head of smallstock.

The district had sixteen *dihan birlishik* (peasant farmer associations) in 1999. In line with national policies promoting the cultivation of strategic crops – wheat and cotton – the area of irrigated land under cereals (mainly wheat) has increased in this district by a third since 1996. Cotton is not grown in this district. The area planted to fodder crops (lucerne, and maize for silage) has declined by half in the same period, while annual forage grasses are no longer planted at all. From 1996 to 1998, therefore, the area of cropped land for growing fodder had declined from 32 to 10 per cent. Increases in the area of wheat grown in Gok Tepe district are shown in Table 11.7. Wheat has replaced vegetables and fruit. However, despite the yield figures shown in Table 11.7, it was stated by a senior administrator that the general trend in wheat yields was downwards. The administration planned to reduce the area of wheat to 10,500 ha in the year 2000. The land released will be used to grow fodder, i.e. alfalfa and barley for cattle, since a severe shortage of

Table 11.7 Area and yield of wheat in Gok Tepe district

Year	Area (ha)	Yield (t/ha)
1997	9,000	1.73
1998	10,000	1.95
1999	13,000	n/a

fodder is recognised. It was also planned to introduce a crop rotation of alfalfa, barley, maize, wheat and perhaps cotton.

Broadly, there are four agro-ecological zones used for livestock in the district. Detailed information is given in Chapter 12, and only a brief summary is presented here. Rainfall declines from 230 mm/year to 150 mm/year from south to north. The sample of livestock-rearing households was selected according to residence in the following four zones.

The first zone, in the south, is the foothills of the Kopet Dag mountains. This area is used by residents of the farm association centres adjacent to the Kara Kum canal for grazing animals. As rainfall is a little higher in this zone, pastures are more productive.

The second zone is the irrigated zone, with water being supplied from the Kara Kum canal. This zone, running roughly south-east to north-west, is used primarily for growing crops of cereals, vegetables, grapes and fodder. Most of the dairy cows in the state sector are kept in this zone. Small numbers of sheep are kept, mainly private sheep and a few owned by the associations for slaughter. The administrative centres of the farm associations are in this zone.

The third zone is a complex of sand and salt deserts. Unlike in Yerbent association, there is little sand/clay desert in Gok Tepe district.

The fourth zone in the north of the district is sand desert, with lower rainfall than the central desert. These two zones are used exclusively for keeping livestock, predominantly on an extensive basis. Due to the low rainfall, only melons can be cultivated, and these only in good rainfall years.

Livestock in Gok Tepe district

Changes in the numbers of small ruminants (sheep and goats) and camels in the district since 1988 are given in Table 11.8. In addition to the livestock shown in Table 11.8, there are approximately 4,500 state-owned cattle in the district, nearly all of which are kept in the canal zone.

While the total numbers of small ruminants and camels appears to have remained relatively stable over the past decade, there has been a large shift in the pattern of ownership, with a significant decline in the numbers in the state sector accompanied by an increase in privately owned livestock (Figure 11.1). In the Soviet era there was a limit to the number of private animals that individuals could own – 20 sheep and goats, one cow and calf, and one camel and calf. However, since independence these limits have been removed. The increase in private ownership has been particularly marked for sheep and goats, whose numbers have increased more than a two and a half-fold over the ten-year period. However, these numbers must be treated with extreme caution.

The number of livestock in the state sector controlled by the farm associations is probably accurate. However, in the private sector there is no census, only a survey of a sample of flocks and, as already explained, people are wary about divulging the size of their private flocks and so probably under-report the number of animals.

Table 11.8 Numbers of sheep, goats and camels in Gok Tepe district, 1988–9

Year	Sheep and goats			Camels		
	State	Private	Total	State	Private	Total
1988	62,073	30,938	93, 011	5,162	1,475	6,637
1989	66,975	39,743	106,718	5,265	1,428	6,693
1990	67,676	40,665	108,341	6,319	1,246	7,565
1991	66,270	37,308	103,578	5,333	1,426	6,759
1992	62,848	33,310	96,158	5,457	1,278	6,735
1993	66,803	44,762	111,565	5,833	1,593	7,426
1994	61,486	55,297	116,783	5,801	1,782	7,583
1995	57,541	63,134	120,675	5,796	2,150	7,946
1996	47,329	62,976	110,305	5,450	2,123	7,573
1997	32,936	65,110	98,046	4,990	2,358	7,348
1998	29,617	61,497	91,114	4,962	2,067	7,029
1999	26,294	85,947	112,241	4,691	2,132	6,823

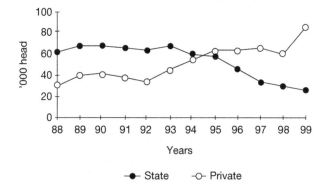

Figure 11.1 Changes in the ownership of livestock between state and private sectors, Gok Tepe district, 1988–99.

On the other hand, it is possible that local officials inflate the numbers of private livestock, in view of the national target for the number of animals in the state sector to increase by 5 per cent each year. As this target is clearly not being met, local officials may be trying to compensate for this by claiming that there are more private (but uncounted) livestock than is the case. The large increase in the numbers of private sheep and goats in 1999 may be a consequence of shepherds who lease livestock receiving a share of the offspring from the flocks, but may also reflect an overestimation of private animals.

Most of the sheep are kept in the desert, which both local officials and shepherds agree is a more suitable environment than the canal zone for raising livestock. Only some 5,000–6,000 (about a quarter) of the state sheep are kept in the canal zone. A few sheep are kept in the foothills area, but the numbers are not clear. According to the livestock specialist of Gok Tepe district, only three out of sixteen associations keep sheep or goats in the foothills area, due to the limited area of grazing land and its rugged terrain.

Pasture land of Gok Tepe

Land in the district is allocated to each of the sixteen associations, while irrigated land in the associations is now leased to individuals, in accordance with national policy. Areas of desert pastures were nominally allocated to associations, with typically each having three or four separate areas of desert land, although not all associations have land in the desert. However, these areas are not exclusively grazed by animals from the nominal association (see Chapter 9). It is common for livestock from several associations to be kept in one desert village, even being tended by different members of the same family.

Livestock management in the desert of Gok Tepe district

Most shepherds are descended from families which have been in the desert for generations. Their ancestors were nomadic, although the distances travelled were not far, perhaps a maximum distance of 100 km. In the northern part of the desert, there were many villages where camels are more important than sheep, because of the lower rainfall than further south. Lower rainfall results in more sparse vegetation and a higher salt content in the water of many wells, which can more readily be tolerated by camels than sheep. However, every family owned private sheep, goats and camels.

Sheep

In 1998 all the associations' sheep in Gok Tepe district were leased to shepherds, usually to two or more brothers, or groups composed of a father with adult sons. In that year most of the leasing arrangements were such that the shepherds kept 40 per cent of the lambs, with a target lambing percentage of 90, and the associations were supposed to supply fodder. In 1999 the contracts were on a 50 per cent basis, but the shepherds were responsible for providing all the inputs.

In the northern desert areas of Gok Tepe long-distance migration of sheep is no longer commonly practised, in contrast to Yerbent association where shepherds linked to the association still undertake a winter migration. The sheep are either grazed within 10 km of the settlement all year round or moved up to 20 km to another area centred on a different well during winter. In summer the sheep were watered every two days depending on the temperature and every three or four days in winter (as low as once per week in winter if the humidity was high). Most sheep were not fed fodder in winter, but relied on pasture all year round, with only weak or old sheep which had lost their teeth being fed fodder. Some shepherds also fed some fodder to old and pregnant sheep if there was extended snow cover. The only time when all sheep received fodder was in severe snowy winters, but this had not happened since independence.

Some wells are too salty for sheep, especially young lambs, and the water has to be mixed with water from elsewhere, usually one part good water with four parts salty water. Sometimes water is transported from the pipe at Kashatakyr, but if the pipeline is not running it has to be transported from Yerbent. Some families have a tanker that they use for transporting water.

Camels

It is common for stock from one association to be looked after by different members of the same family. For example, one family in a desert village comprised the father, who was retired, and five sons, all but the youngest being married. The five sons worked as follows (in decreasing order of age):

1   Camel herder for Hourmant association
2   Looks after private sheep and goats for whole family
3   Camel herder (thirty-two camels) for Kelejar association
4   Camel herder (twenty-five camels) for Kelejar association
5   Camel herder (ninety-six camels) for Gok Tepe association

Sometimes, some of the brothers herded their camels together, even although they belonged to different associations.

Most association camels are leased to the herders. For example, Gok Tepe association has fifteen camel herds of which fourteen are leased out. The unleased herd consists of male camels and the herder is paid a salary. The herders of the leased camel herds are set a target calving percentage of 45 per cent (she-camels usually calve every second year). The contracts state that the herders keep half the output of young camels. If the target is not achieved the association's share has to be made up from the herder's share but if the target is exceeded, the herder keeps the excess. The herder keeps the wool, although at present it is worth very little. The herder has to provide inputs such as fodder, although if it is available the association may supply it at cost price.

Private livestock

All households interviewed in each of the four zones keep private livestock. The numbers vary considerably with an individual having ten to over one hundred adult sheep and goats (the number of goats is usually about half the number of sheep). Often the private sheep from several family members are herded together with the association sheep or, if there are no association sheep, the private sheep of several family members may be tended as one flock. Usually each family has a few private camels to provide milk, with numbers varying from two to fifteen.

Some shepherds are paid to keep livestock for others, under the *chekene* system described in Chapter 9, charging about 3,000 manat per sheep per month. Some shepherds who also had association sheep did not consider this to be enough, because they had enough work looking after those sheep and their private sheep, but did so because the sheep belonged to relatives.

Currently there is no restriction on where individuals can graze their livestock and shepherds are free to use the land allocated to the associations.

Average numbers of private livestock from the sample of households interviewed in each of the four zones is given in Table 11.9. The largest private flocks were in the northern desert zone (as is the case in neighbouring Yerbent), with a household average of ninety-five sheep/goats and twelve camels. Flock size became progressively smaller moving from the central desert to the canal zone and the foothills. In the foothill zone none of the surveyed families kept camels, but kept cattle instead.

Sheep and goat productivity in terms of lambing percentage (overall average 97 per cent), was highest in the canal zone, reflecting the better quality feed from crop residues, alfalfa, etc., available in the irrigated areas. In the northern zone of the central Kara Kum desert, lambing percentage was only 83 per cent.

Fodder

None of the households surveyed routinely fed supplementary fodder to their livestock in winter. General feeding was only carried out in winter when there was heavy snowfall, which only occurs every eight to ten years and has not occurred since independence, or to sick or very thin animals. In some cases supplementary feeding is provided to animals which are being fattened to be slaughtered for meat.

In the desert areas fodder was bought, usually from traders who visit the settlements and buy wool or camel wool from shepherds. In the canal and foothill

Table 11.9 Average number of private livestock per household in sample, Gok Tepe district

Zone	Number of h/holds	Sheep and goats				Camels (or cattle)	
		Average per h/hold	Ewes	Lambs born 1999	Lambing percent.	Total per h/hold	She-camels
Northern desert	15	95	37	31	84	12	6
Central desert	15	81	33	29	88	6	3
Canal	10	73	32	31	97	5	2
Foothills	5	69	31	27	87	3*	1*

Note
* Indicates cattle instead of camels.

zones shepherds usually make hay themselves. Typical prices for fodder bought by shepherds in the desert in 1999 were:

Bran: 350–400 manat/kg
Barley: 500–600 manat/kg (not often available)
Yandak (dried camel thorn): 110–75 manat/kg

Further information on the costs of fodder provision is provided in Chapter 12.

Worsening social conditions in the desert

The level of education is low. The majority of shepherds were illiterate but numerate, and knew exactly how many animals they had. Their children no longer attended school, as smaller schools are no longer staffed. There are still schools in some villages, but often it is too far from outlying villages for children to walk each day. For example, at one village the nearest school was 7 km away, too far, it was claimed, to travel to school every day.

Livestock marketing

Livestock is sold when the family needs cash. Animals are usually transported to the bazaar in Ashgabat, using transport provided either by a family member who has a truck, or a truck-owner who charges for the transportation of the stock. Transport costs and market prices are shown in Table 11.10.

Conclusions

The pace and thus the impact of reforms in Turkmenistan has undoubtedly been different to that in Kazakstan. The reforms in Turkmenistan are mainly driven by Presidential Decree, with the details of implementation being left to different ministries and organisations like Turkmen Mallory. In many cases the policies are not well defined and therefore can be interpreted in different ways. Because of this, the impact of policy changes can be difficult to predict, especially because there is continual change in the way in which policies are implemented. The changes in the livestock leasing system, described in Chapter 9, have meant

Table 11.10 Marketing costs and returns, from Gok Tepe desert to Ashgabat, spring 1999

Type of animal sold	Cost of transport to Ashgabat (manat)	Price received per animal sold (manat)
Sheep	9,000	120,000–200,000
Female camel and calf	100,000	2,700,000–3,000,000

an unstable economic, institutional and policy environment. Under these conditions it is very difficult for producers to plan their production systems; also it makes the introduction of new technologies more difficult.

The policy of increasing wheat production to ensure self-sufficiency in wheat flour for bread-making has restricted the amount of land available for fodder production. Although the amount of fodder needed in most winters by sheep in the desert is very small, every ten years or so there is a very severe winter with heavy snowfalls. In the past under the Soviet system there was always sufficient fodder available, along with the means of distribution, to ensure that most livestock survived. There has not been a severe winter in the study area since Turkmenistan became independent. A bad winter with heavy snowfalls which prevent livestock from grazing, coupled with the shortage of fodder, would result in a huge death rate amongst the sheep flocks and have devastating consequences for the size of the national sheep flock. Such a winter occurred in 1999 in Mary *velayat*, where thousands of animals were lost.

Some commentators within Turkmenistan have argued that land should be privatised on the grounds that privatisation provides an incentive for producers to care for the land in a sustainable way. However, this option would be difficult to implement under the current political and economic uncertainties. If shepherds are forced to sell the output from their leased animals back to the associations at a fixed price, they will not be willing to invest in land and other resources such as wells. Furthermore, at present stock are moved, albeit relatively small distances, in the different seasons of the year. These movements allow the stock to take advantage of the seasonal variation in the productivity and availability of different types of vegetation and to exploit water resources of varying quality. Although the allocation of discrete areas of land to households may be a suitable land tenure system for cropping land, it may not be suitable for large tracts of desert. The subdivision of such land into small areas may prevent the sustainable exploitation of the desert's different resources, by preventing the traditional movement of stock.

The livestock sector of Turkmenistan can undoubtedly contribute more to the national economy, and is the only way in which vast areas of desert – which cover most of the country – can be used productively. The shepherds of Turkmenistan have shown over the past two or three years that they can respond very rapidly to changes in the political and economic environment. The shepherds have modified their production systems accordingly, to produce meat, wool and fibre from these resources. There is an urgent need for stability in the institutional arrangements which affect livestock production to allow them to play their full part in the post-Soviet development of Turkmenistan.

12

THE COSTS AND RETURNS
OF CHANGE

Profiles of production and consumption by pastoralists

Ogultach Soyunova

Introduction

Restructuring of the livestock sector in Turkmenistan has already had marked economic impacts on the pastoralists. This chapter shows that these impacts vary by location and by the wealth status of pastoral households. Those living deep in the interior of the desert have distinctive constraints as well as opportunities compared to those living nearer the centres of population, irrigated land and major transport routes. There is now a disparity in the size of private flocks and herds among pastoral households, since the ceiling was lifted on the number of animals which a family may privately own. The margins of costs and returns from livestock now differ with the scale of operation, as some households have been able to accumulate more animals since reforms introduced in the 1990s. Differences in livestock wealth are also reflected in consumption patterns at the household level.

There are also certain general trends proceeding from reform that apply to all pastoral households, regardless of their location or wealth status. Principal among these is that all pastoral households must now rely on selling livestock to obtain their basic needs. Income through employment on state farms has been reduced or eliminated, resulting in much economic insecurity as well as an absolute decline of income in numerous cases. For many pastoralists, obtaining winter fodder for their livestock has become a major difficulty, while for all there is a growing gap in the terms of trade between livestock and food prices.

Methods of study

Two study areas, described in Chapters 9 and 11, were surveyed by informal interviews and formal questionnaires to respondents.

In the first study area of Yerbent, a large farm association, forty-seven shepherds with private flocks or with animals leased from the farm association answered

228

a questionnaire in the late spring of 1998. There were forty-four responses on live-stock management as three respondents had no sheep of their own. Of interest was whether any economic stratification could be observed among shepherd families, after privatisation and reform policies had taken effect. Therefore this survey investigated whether livestock management, income and expenditures differed according to the livestock wealth of shepherd families. Shepherd households were selected from four different settlements within the Yerbent farm area, using the same sample as reported in Chapter 11. Responses were grouped into four classes according to the size of privately owned flocks; those with twenty-five or less livestock, those with between twenty-five and fifty animals, a third group with fifty to one hundred animals and a fourth group with one hundred or more animals.

Within the second study area of Gok Tepe *etrap* (district), forty-six shepherd families were interviewed by questionnaires in the autumn of 1999. In this case, it was decided to investigate whether there was any clear geographical differentiation between shepherd families in different agro-ecological zones, described in Chapter 10. The differences considered were in shepherd family incomes, expenditures and livestock management. Sampled households were selected within four zones running from north to south within Gok Tepe district. Responses were then grouped for analysis into four location categories of northern desert, central desert, canal zone and the southern foothills.

Livestock management, income and expenditures in Yerbent

The association had 50,000 sheep and 2,300 camels in 1998, managed by some 400 shepherding families in five farms. In 1998 when the survey was conducted, some of these shepherds had begun to lease sheep and camels from the association, while others were still employees of the association (see Chapter 9). Most households in Yerbent area, whether members of the association or not, have their own live-stock, especially now that reforms have allowed shepherds to lease state animals and keep some of their offspring. The sample of households was grouped according to private flock size as noted above, in order to observe any economic differentiation between shepherding households.

Flock size and offtake patterns

Some differences between households based on flock size are shown in Table 12.1 and Figures 12.1 and 12.2. Seven households of the first group owning less than twenty-five sheep had less than five sheep. All households rely on selling animals, with a rate of between 15 and 24 per cent of the current sheep flocks sold in the previous year. Smaller flock-owners sell proportionally more of their animals – both smallstock and large animals – than do bigger owners. Larger flock-owners, being relatively better-off, have less need to sell and are trying to accumulate greater numbers of animals.

Table 12.1 Annual production, income and expenditure from livestock, sample
households in Yerbent, 1998

Households with privately owned smallstock n = 44	Less than 25 head n = 21	25–50 head n = 7	50–100 head n = 9	More than 100 head n = 7
Mean number of livestock				
• sheep or goats	6	19	49	133
• camels	3	5	15	16
Mean head/amount sold				
• sheep or goats	2.7	8.2	11.6	32
• camels	1	1	3.3	4.8
• wool (kg)	29	40	173	339
• meat (kg)	86	nil	250	214
Cash income from livestock products sold ('000 manat)	2,246	2,803	6,194	12,845
Value of in-kind income from livestock* ('000 manat)	1,101	1,759	2,406	2,968
Cash expenses on livestock ('000 manat)	2,485	1,837	7,976	6,990

Note
* Valued at current market prices.

The sale of camels is as important to household finances as the sale of smallstock, providing about half the cash income of flock-owners in all flock size categories. Camels are worth from five to ten times the market value of smallstock, depending on the type of camel sold. Cash from selling camels is proportionally more important to smaller flock-owners who cannot afford to sell many smallstock but must meet their minimum requirements from the market (Figure 12.1).

The imputed cash value of different animal products consumed or used by households with different flock sizes is shown in Figure 12.2. For the smaller flock owners the subsistence value of meat is greater than the cash value of smallstock sales. However, cash rather than subsistence is the major contributor to all households' income from livestock when income from selling camels, meat and wool is included. This is shown by comparing Figures 12.1 and 12.2. All households now have to rely on selling animals to obtain wheat flour through private market channels. Informal interviews with pastoralists revealed that the main use of cash from livestock sales is to buy flour, followed by tea, sugar, clothing and children's necessities.

By comparing Figures 12.1 and 12.2 it is clear that larger flock-owners are already considerably commercialised, despite a lower annual sales rate than households with smaller flocks, as they derive a much greater proportion of their total livestock income from cash sales.

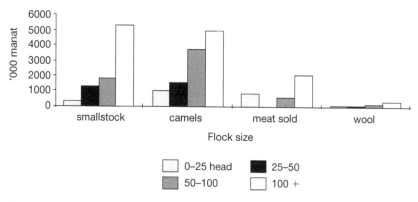

Figure 12.1 Annual household cash income from livestock, Yerbent, 1998.

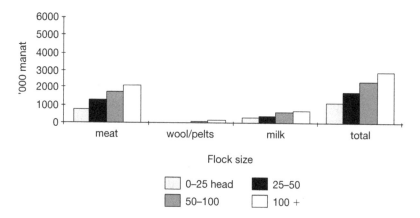

Figure 12.2 Annual household in-kind income from livestock, Yerbent, 1998.

The sale of wool, wool items and pelts brings negligible income to households in all economic classes, as indicated in Figure 12.1. This is a major change from the Soviet period when the cash value of wool was much higher, and the wool shorn from one sheep could buy a sack of flour sufficient to feed a family for a month. Now this amount of wool would buy only a loaf of bread. By the late 1990s, wool had so little market value that it was sometimes burned, and the cost of shearing a sheep was not met by the price of its wool. This problem is further discussed in Chapter 13. In interviews, pastoralists stressed that the deep decline in terms of trade for wool to other commodities was one of the major economic difficulties they now faced. Nevertheless, the subsistence value of wool is still important to families, as coarse sheep wool is made into felt floor mats while camel hair is used for stuffing quilts. Demand for these items in the markets is not sufficiently high to bring a good income, however.

Household consumption of livestock products

A noticeable pattern is that household consumption of meat and milk rises in direct relation to flock size (Figure 12.3). Thus, members of better-off households appear to consume twice the amount of meat – 66 kg per year per person – than those with the smallest flocks. The regional standard for meat consumption is 60 kg per person per year or 164 g per day. However, some meat consumption may be by visitors, as by custom, better-off households are expected to entertain guests by slaughtering an animal.

The value of milk to household economies is greater than suggested by Figure 12.3 since the value ascribed to milk is based on the urban market price, which is much lower than for meat by weight. In terms of quantity, milk contributes from 62 litres up to 104 litres per person per year from households with the least and most animals. This means that even the poorest households consume 170 ml of milk per person per day. Most milk consumed is from camels, as sheep are very rarely milked. This finding underscores the great importance of camel milk as the main source of protein in the desert diet.

Expenditures

The principal expenditures on animals among all households were for capital costs involving water points, vehicles and lastly shelters for animals (Figure 12.4). The main other expenses were recurrent costs for purchasing winter fodder and fuel, the latter needed for motorbikes and vehicles used to herd flocks, transport of fodder from other areas as well as transporting animals to markets. The cost of hiring vehicles is also important. A negligible amount of funds are spent on veterinary inputs, as the households surveyed were all in the desert area, where animal health status is generally much better than around the settled canal zone, as discussed below in the section on Gok Tepe district.

A clear pattern is that the larger the flock, the lower the expenses on fodder per head (Figure 12.5). Households with the most sheep spend on average half the

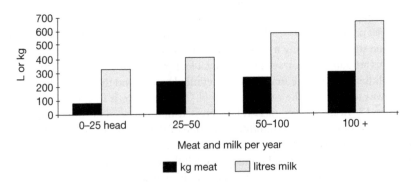

Figure 12.3 Annual consumption of meat and milk by households, Yerbent, 1998.

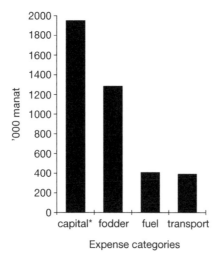

Figure 12.4 Annual expenditures by households on livestock, Yerbent, 1998.

Note
* Capital items are mainly water pumps, vehicles and winter shelters.

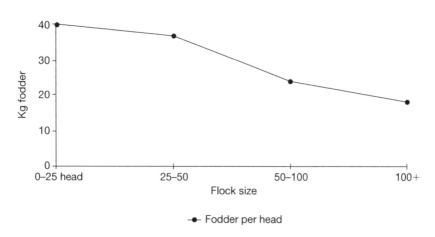

Figure 12.5 Annual amount of fodder per head, households in Yerbent, 1998.

amount per sheep on fodder, compared to households with the fewest sheep. This economy of scale is achieved mainly because owners of larger flocks are more mobile in seeking out natural pastures. Larger flocks are therefore less reliant on supplementary fodder. Smaller flocks tend to be grazed around the perimeter of settlements most of the year, as noted in Chapter 9.

Overall, smaller flocks seem to require greater investments per head of livestock, as suggested in Figure 12.6. A large part of expenses is in the cost of providing

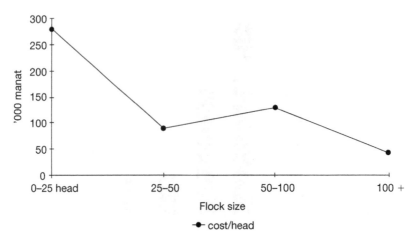

Figure 12.6 Annual household expenditures per head sheep, Yerbent, 1998.

supplementary fodder, as has been indicated above. The pattern noted here is common in other pastoral regions of the world, where very small herd-owners must devote proportionally more capital to their flocks. Small herd-owners cannot achieve the economies of scale that larger herd-owners can, for example in the cost of renting a vehicle or maintaining a water point which will equally serve for a few or many animals.

Returns from livestock

Net income from livestock, after deducting expenses of watering, feeding and transport, is ten times higher for households with the largest as compared to the smallest flocks (Figure 12.7). Households with the most animals do not, however, contain more people, having an average of four as compared to five persons in the poorest households. However, while the finding that net income increases with number of animals owned is quite predictable, living standards even among households with more animals is now quite low compared to the recent past, while households with very few animals are in an even more precarious state with regards to obtaining basic food needs.

Despite, or because of their poverty, the smallest-scale livestock-owners derive the most value *per head* from their animals, as Figure 12.8 shows. Each animal therefore yields more for the owners of the smallest flocks. But although smaller flocks are more economically efficient, they cannot provide sufficient income and security to their owners under the present-day uncertain conditions.

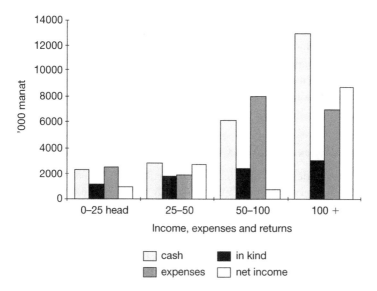

Figure 12.7 Returns in cash and kind from livestock, Yerbent, 1998.

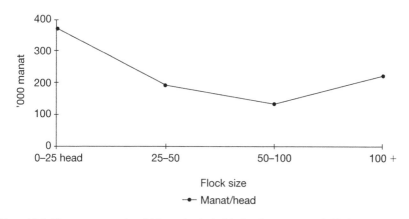

Figure 12.8 Net returns per head/sheep (cash, in-kind, minus expenses), Yerbent, 1998.

Livestock management, income and expenditures in Gok Tepe

Different impacts of the reforms were sought in the survey on livestock-owning households in Gok Tepe district, compared to the survey in Yerbent area. The Gok Tepe survey focused on variations in livestock productive output between households located in different ecological zones.

Among the forty-six sampled shepherd households, fifteen lived in the northern desert, at a total of five different hamlets; sixteen households were selected from three hamlets in the central desert zone and ten families were selected from two farm association centres in small towns along the Kara Kum canal. Finally, five families were selected who grazed their animals in the foothills of the Kopet Dag mountains in the south of the district, based in two farm association centres along the canal.

Types of livestock ownership and management

The majority (81 per cent) of households in the sample had leased livestock from an association. Very few households have only private animals, while a higher proportion manage sheep for others, hired as shepherds. Among the fifteen sampled families in the northern desert zone, animals – mostly camels – were leased from six different associations. In the central desert zone, ten of the sixteen families had leased animals from five different associations. In the canal zone sample of ten families, four leased livestock from two associations, three families had *chekene* (hired-in) flocks, while the remaining three families had all three types of flocks; leased, private and *chekene*. Four of the five sample families in the foothill zone had leased flocks from one association.

Several general patterns can be seen from Tables 12.2 and 12.3. The far northern interior of the desert is used for raising leased camels and privately-owned sheep, while the central desert, canal and foothill zones are used by all types of flocks – leased, private and hired (*chekene*). Table 12.3 shows some variation in flock structure and output within the different ecological zones. Northern desert private sheep flocks are smaller than those in the central and canal zones, as camels

Table 12.2 Livestock management by sample households, Gok Tepe, 1999

Livestock managed by households	Northern desert n = 15	Central desert n = 16	Canal zone n = 10	Foothills n = 5	Total n = 46
Leased and private	11 leased camel herds	1 leased camel herd	4 leased sheep flocks	4 leased sheep flocks	34 (74%)
	4 leased sheep flocks	10 leased sheep flocks			
Private only	nil	2 sheep flocks	nil	1 sheep flock	3 (6.5%)
Leased, hired in (*chekene*) and private	nil	nil	3 sheep flocks	nil	3 (6.5%)
Hired in (*chekene*) and private	nil	3 sheep flocks	3 sheep flocks	nil	6 (13%)

Table 12.3 Flock size and lamb output by zone and management system, Gok Tepe, 1999

Zones and management type	Number of flocks in group n = 65	Average number of sheep per flock	Percentage of ewes per flock	Number of lambs per 100 ewes
Northern desert leasing	4	383	72	91
Northern desert private	14	47	58	90
Central desert leasing	8	296	66	90
Central desert private	14	64	55	103
Central desert hired-in	4	261	79	98
Canal zone leasing	4	280	64	88
Canal zone private	8	124	60	77
Canal zone hired-in	3	182	78	84
Foothills leasing	4	558	47	99
Foothills private	2	35	73	92

are the preferred species in the north. The largest average number of private sheep per flock are kept in the canal zone, where grazing land is least available but other sources of income and access to crop residues allow families to keep more sheep. Among leased flocks, those kept in the foothills are biggest, and have the highest proportion of male animals; these flocks contain many animals being kept for the market.

The different types of flocks have a variable proportion of ewes. Overall, a higher rate of ewes is found in leased and hired flocks, with the exception of the foothill zone.

Livestock productivity by ecological zone

Flock productivity is measured by the number of lambs born per hundred ewes, which in Gok Tepe district are all of the Saraja breed. The district government authorities still set lambing targets for shepherds to reach with their leased flocks. In 1998/9 the target was 93 lambs per 100 mature ewes. This target was only met among foothill zone flocks, however.

An interesting result is the higher lambing rate for privately-managed flocks and the gap between rates for leased and private flocks in the central desert zone (Table 12.3). Officials and shepherds consider the desert as the optimal area for raising sheep. This result seems to reinforce that view.

Income from livestock

The amount and sources of annual income in cash from livestock varies by zone (Figure 12.9). Households in the northern desert derive about half their cash income from selling camels, selling just over three camels per household on average in 1998/9 at an average price of 1.7 million manat each. Although households in the

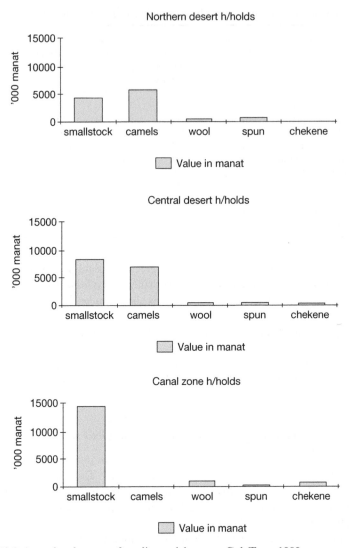

Figure 12.9 Annual cash returns from livestock by zone, Gok Tepe, 1999.

central desert zone sold slightly more camels on average, their main source of income was the sale of smallstock which provided half their annual cash income. Central desert households sold nearly twice the number of smallstock per household than did northern households and also have a higher offtake rate. Income from the sale of smallstock is highest in the canal zone, where this provides 80 per cent of total cash income from livestock. These differences are due to the greater accessibility of markets for households in the central and canal zones as compared to the northern desert.

Income from sales of sheep, goats and camels provides the principal source of income for desert families. These families, especially in the central desert area, earn much smaller amounts of cash from selling raw or spun wool and felt mats. With more favourable market access, livestock-keeping families in the canal zone gain higher incomes from selling smallstock, while access to irrigated land allows them to also earn income from growing and selling crops. These findings reveal the extent to which desert families are now dependent on livestock marketing for their income.

Expenditures

Expenditures on livestock are distinctly differentiated by zone. In the desert pastures, the main expenditures are for transport related to watering, grazing and marketing livestock. Funds must be spent on fuel for water carriers, repairing or acquiring water pumps, purchase and repair of motorbikes used for shepherding. Lastly, desert shepherds expend more on fuel for motorbikes and for trucks used to transport animals to the city markets. In contrast, within the canal and foothill zones the bulk of livestock-related expenditures is on veterinary medicines. In these areas, livestock parasites and diseases are commonly transmitted through the standing water from irrigation canals (see Chapter 9). Annual expenditures on veterinary drugs by families in the canal and foothill zones averages 1,500 manat per livestock head, a cost not incurred by those families keeping livestock in the desert.

Conclusions

A number of economic differences have developed between pastoralist families, as a consequence of the recent reforms. Families with very few animals must depend on them for subsistence – meat and milk – while people with more livestock are able to gain a net cash income over and above their basic requirements. Subsistence income from animals remains remarkably constant across families with different levels of animal wealth, but income from sales rises steadily with flock size. The transfer of livestock ownership from the state to private families is gradually allowing some families to accumulate animals and to manage these on a semi-commercial basis. The marketing channels which now allow producers to sell on open markets (Chapter 13) offer better-off families a chance to profit from their livestock.

Most families who depend on livestock still have very few animals, however. The milk from these animals – camels and cattle – forms a very important part of family diets. Animals must be sold regularly to supply wheat flour. Expenses on livestock are proportionately higher than for better-off families, due to diseconomies of scale.

Geographical location also plays a role in how well pastoralist families can manage their animals under the new conditions of privatisation. The advantage of living in the interior of the desert is generally that animals are healthier and have more access to pastures. This reduces recurrent expenditures on veterinary medicines and fodder. The desert is particularly suited to keeping camels, which provide a source of income when sold. However, desert families have to spend more on transporting animals to markets, to which they have less ready access, compared to pastoralists in and around the Kara Kum canal zone. For the latter, access to fodder from crop residues is also an advantage, but there are costs in terms of animal health.

Reforms will continue to affect every aspect of livestock management for pastoralist families. Some of the economic differences now emerging between families, based on livestock wealth and geographical location, can be expected to deepen in the future.

13

PRIVATISATION OF LIVESTOCK MARKETING IN TURKMENISTAN[1]

Carol Kerven

Introduction

One of the most important outcomes of the transition from socialist to market economies has been the liberalisation of state-controlled exchange and distribution channels. This process has occurred in the livestock sector of Turkmenistan over the second part of the 1990s. The state has not yet surrendered complete control over this sector, but there are signs that a vigorous private sector is taking over the purchase, processing, retail and export of livestock products. In contrast, state marketing systems are now generally dysfunctional and not commercially oriented.

Livestock producers – the pastoralists – are not, however, benefiting as rapidly from the newly liberalised marketing structures. Producers currently experience poor terms of trade for livestock products in relation to inflated prices for basic commodities. Another reason is that the state organisations are still able to intervene in setting exchange values for livestock products which deprive producers of the full market return. A further reason is both state and private marketing organisations have not been to able to successfully enter world markets for some key livestock commodities produced in Turkmenistan. Thus producers are no longer able to realise a viable return on some of their livestock outputs, which remain unsold and unused.

In the post-socialist economy, Turkmen pastoralists have become very dependent on selling live animals to support their basic subsistence. However, unlike in Kazakstan, the rise of a distinct group of commercially minded pastoralists is not yet widespread.

1 The material on which this chapter is based was partially obtained during a study for the Global Livestock Collaborative Research Support Programme, 'Integrated Tools for Livestock Development Rangeland Conservation in Central Asia', by the University of California, Davis.

This chapter discusses some of these changes in marketing systems resulting from policy shifts since independence.

Policy changes affecting livestock marketing

Marketing of livestock and animal products has been partially privatised as a result of policy initiatives in Turkmenistan since Independence. Two key legal changes have promoted this change; the lifting of restrictions on the number of animals that may be privately owned, and second, the introduction of a leasing system for state-owned livestock. Individual families can now accumulate livestock through purchase, inheritance, natural reproduction and as their share of offspring from animals leased from the former state farms. One result is that the national proportion of privately owned animals has been increasing while that of state animals declines, and the majority of animals are now privately owned. Figure 13.1 shows the proportions in public and private ownership for the region in which the study was carried out. Thus, the main flow of animals and products to the market is now through non-state channels. National statistics on numbers and ownership of animals sold are not collected, however.

Production of meat for domestic consumption is an important government policy issue, as the country is not self-sufficient in meat. At the end of the Soviet period, Turkmenistan relied on importing some fifty thousand tonnes of meat mostly from the USSR. This amount increased by 30 per cent in the first five years after Independence, as state meat production plants ceased to operate (TACIS 1997). Meat output declined nationally, especially from fodder-dependent cattle and pig-raising state production units where provision of feed became problematic. During 1993–5 meat output from cattle and pigs declined by a half and three-quarters respectively. By the mid-1990s, government emphasis was placed on increasing the national smallstock population, which can be fed on natural pastures. This is discussed further in Chapter 11.

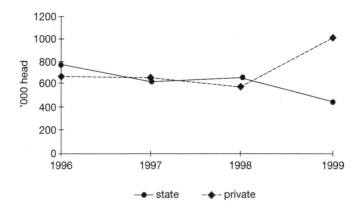

Figure 13.1 Number of smallstock in Ahal region, 1996–9.

State and private marketing channels now exist in parallel. State livestock farms have been converted into farmers' associations (*dihan birlishik*) whose livestock are leased to families on an annual contract. This system is called *arienda*, and is described in Chapter 9. The contracts specify that a share of the offspring and products in the form of pelts and wool must be returned to the association's administration, to be sold or consumed to cover their overhead costs. The state organisation Turkmen Mallory also collects wool and pelts from individual associations for sale to commercial buyers. Meat, wool and pelt processing factories have also been partially privatised along the leasing system, with lessees returning a share of their revenue to the state. Another indication of privatisation is that private traders have been sub-contracted by the government to sell state-owned animals collected from associations, paying the sales income to the government and retaining any profit on transactions.

Private marketing channels are either directly from producer to consumer, via the urban livestock markets, or through traders who buy and sell in bulk. In the study areas, most producers interviewed sold directly to the markets in Ashgabat or along the settled canal zone. Traders in livestock, wool and pelts do operate within the study area, but these traders tend to buy more from remoter areas further from Ashgabat, where individual producers have greater difficulty reaching the market. The study area is relatively accessible to Ashgabat and the settled canal zone, given the number of vehicles available in the desert and the low cost of fuel.

Methods of study

The study was carried out in four different areas, during spring 1999.

- The central Kara Kum desert in Gok Tepe *etrap* (district) and Yerbent farm association, which are described in Chapters 9, 10 and 11.
- The irrigated settled agricultural zone within Gok Tepe district around the Kara Kum canal which runs from east to west along the foothills in the southern part of the country.
- The town of Gok Tepe, along the main road west of Ashgabat, at the livestock market and administrative headquarters (*Hakimlik*) of the district.
- Ashgabat city, at two livestock markets and in offices.

Further information was collected in Ashgabat and Mary cities in spring 2000.

Interviews were carried out with thirty shepherding families, women carpet-makers, farm association professional staff and directors, livestock traders at three urban markets, urban commercial companies dealing in wool and pelts, senior staff of the Ministry of Livestock (Turkmen Mallory) and European Union project staff involved in a livestock project. Respondents were selected to include representatives of most groups involved in livestock production and marketing of wool, animals and other animal products.

The method of interviewing was a semi-structured open-ended checklist of questions. Field work in the desert and canal zone involved staying with local residents. This provided opportunities for more informal discussions on the topics of the study as well as observations on livestock management practices and the role of livestock in household economies.

Contribution of livestock to household economies

All families in the desert area and most within the canal zone settlements keep livestock. Within the desert villages there are three types of families; those with a wage job (mainly teachers and some government employees); those managing leased animals from an association, and lastly, those managing their own private animals and shepherding for others (a system called *chekene*). There is remarkably little difference in the standard of living between these types of families. The main differences are that richer families, who have more livestock, eat more livestock products, have a truck or jeep and perhaps a motorised pump for a well.

For families without wage employment or access to irrigated land, the sale of live animals is their major source of cash income. This is the case for most families living in the desert, where the majority of animals are kept. The principle commodity for which animals are exchanged through the market is wheat flour for making bread. Livestock marketing for desert families is largely in order to meet dietary caloric requirements. Being unable to grow any staple crops in the desert, families derive their food needs by selling animals. Until the middle of the twentieth century, prior to Soviet collectivisation, Turkmen carpets and wool were sent on camel-back from the desert to the cities of Bokhara and Khiva, to trade for wheat and rice.

Income from selling animals and their products is also the means for desert families to obtain other necessities from the market, principally cloth, tea, sugar and cooking oil. Up to one or two generations ago, much clothing was woven from homespun wool, while silk and cotton cloth was also obtained by trading carpets and wool in the cities. Nowadays some families sell, in addition to live animals, wool and felt carpets, wool and knitted garments and some yoghurt, to obtain clothing for children and little luxuries such as sweets. These items are usually sold by women, though the income is not always disclosed to their husbands.

Desert families still heavily depend on animal products for subsistence needs. Camel or cows' milk is a main constituent of meals (see also Chapter 12). Nearly all desert families (and many village families) keep at least one milking camel in the yard. Many also keep a milking cow. Milk products are usually the only other food served apart from bread, at meal times. Meat is eaten only perhaps on a weekly basis, even in better-off households. Data in Chapter 12 shows that meat consumption more than trebles between families with the smallest and largest flock sizes, while milk consumption doubles. This suggests that there is unmet demand for meat and milk within producer families which will be satisfied by better-off families before more milk products are sold.

244

The pattern of sales, purchases and consumption among families underscores the crucial role of each livestock species for coping in the desert. Sheep are sold frequently, from once a month to once every three months, to buy flour. Sheep are consumed at the rate of one to two per month by each family, depending on the number of people in a family and the size of their flock. Sheep wool, which formerly had a high exchange value, is now rarely sold but is still used to make felt carpets, stuff mattresses, and less often now, hand spun into wool for carpets and clothing. Camel's milk is a main source of food, as is cow's milk. Camel wool, as with sheep wool, is no longer sold but is used for stuffing quilts, weaving into rugs or knitting into sweaters. Camels are also a 'savings bank' to be sold whenever a large amount of money is required, typically to buy a vehicle or spare parts, or as the brideprice for a son's marriage. Goats are another source of meat, and goat hair is mainly used for felt rugs.

Shepherding families receive far less exchange value from managing animals now than in the Soviet period. Table 13.1 shows the erosion in purchasing power of livestock and their products and cash income from shepherding, as a ratio of how much grain can be bought for animal commodities sold. This ratio is frequently referred to by older shepherds as the main index of how their standard of living has declined in real terms. Since the reforms of the 1990s, prices for basic foodstuffs and other household needs have risen out of proportion to cash returns from livestock.

One consequence is that the necessity of obtaining sufficient calories for family consumption has altered the balance between the subsistence and exchange value of livestock to producers. As grain provides more calories than livestock products of meat and meat products, shepherding families now have to sell a greater number of animals than in the past, in order to obtain enough food. Animal sales must now make a greater contribution to household economies. Meat is eaten much more rarely than in the past, as it is more calorically advantageous to sell an animal and buy grain with the proceeds. Milk remains a very important contribution to

Table 13.1 Change in value of livestock to grain, Soviet to present period

Livestock and shepherding income	End of Soviet period, in roubles	Equivalent sacks of flour	1999 manat	Equivalent sacks of flour
Sack of flour (50 kg)	13	—	75,000	—
Average sheep price	120–150	9–12	247,000	3.3
Wool of 1 sheep (3 kg)	12	0.9	1,500	0.02
Sheep skin	6.5	0.5	2,500	0.03
Hired shepherd per 100 head/month	100	7.7	250,000	3.3
State farm shepherd salary/month	37	2.8	—	—

subsistence for several reasons: an animal can be milked for some years then sold as meat for cash to buy grain and second, there is a poor market for milk products, as compared to meat.

Whereas in the past, a shepherd family could obtain their basic needs through selling a few animals or some of their wool, current prices do not allow this. Shepherd families who do not own a large number of animals now have difficulty obtaining their basic food requirements, which leads them either to abandon the desert for the irrigated canal zone, or to become hired shepherds, managing other people's animals for a wage (as noted in Chapter 9).

Marketing of private livestock

Live animals – sheep, goats, camels and cattle – are transported from the desert and canal villages to urban markets by their owners, using their own or rented trucks. Transport is readily available and cheap, due to the extremely low cost of fuel. Transport costs average about 2 per cent of an animal's sale value. Livestock owners in small villages cooperate in transporting each other's animals to market. This is in contrast to the study area in Kazakstan, where transport is a significant cost and truck-owners do not transport their neighbours' animals to market even for a fee (Chapter 8).

The largest livestock market in the study area and in the country as a whole is Jigl Luk, on the edge of Ashgabat. There is also a specialised livestock market in Ashgabat, which is supplied by between fifteen to forty traders (*alup satiji*). Other towns in the canal zone study area, notably Bizmain and Gok Tepe, have small weekly markets at which live animals are also sold.

Sellers must pay a government market tax, amounting to 500 Manat per sheep and 2,000 manat per head for cattle. Except at the traders' market, all transactions are conducted in the open. Sellers stand next to their tethered animals, as buyers inspect the animals and negotiate a price. Purchased animals are taken away in private vehicles. Animals are not sold by weight but the degree of fatness in the tail, and general body size is keenly judged by buyers. Table 13.2 gives livestock market prices in spring 1999.

Several patterns are suggested by the prices shown in Table 13.2. Producers selling directly to the markets received lower prices than did traders. This is not only due to traders' mark-ups, but also due to the timing of sales. Prices obtained by producers were in March, several weeks before the important annual Muslim holiday of Kurban Baihram, when families wish to slaughter the best animal possible. Traders hold onto their sale animals until immediately before the holiday, when buyers are willing to pay a premium price. Second, prices for fattened animals are considerably higher than unfattened animals of the same age. The fattened animals are visibly much heavier (no scales are available in the markets) and can obtain nearly three times the price of an unfattened animal of the same age. Third, price differentials between males and females are not immediately obvious from Table 13.2, since price is so closely related to weight. Very few young females are

Table 13.2 Livestock prices by type of animal and seller, March–April 1999

Species, age, sex and breed sold (n)	Ashgabat market, sold by shepherds from desert Manat price	Ashgabat market, sold by traders Manat price	Gok Tepe market, sold by shepherds Manat price
Sheep (n = 16)			
Saraja breed unless noted			
1 yr males (4)	273,000		300,000
1 yr females (2)		310,000	
2 yr male (1)		420,000	
2 yr male Karakul (3)		320,000	300,000
2 yr females (nil)			
3 yr males (2)	240,000		450,000–700,000
3 yr males fattened (2)		650,000	
3 yr female fattened (1)		550,000	
5 yr male (1)		350,000	
Camels (n = 3)			
4 yr female	1.3 million		
10 yr female with calf	2 million		
15 yr female	2.5 million		
Goats (n = 5)			
4 month male			160,000
2 yr female	190,000		
2 yr male	150,000		250,000
3 yr female			230,000

sold in the market, since owners try not to sell breeding stock, while old females whose meat is tougher and less marketable tend to be reserved for home slaughter by producers. Two-thirds of sheep sold are aged two years or less, and 80 per cent are males. Fourth, interviews with sellers revealed that castrated males and the Saraja breed of sheep fetched higher prices than uncastrated males or Karakul sheep of the same weight. This is due to consumer taste preferences.

Overall, the returns from marketing sheep depends on their weight and degree of fatness. Neither age nor sex is considered by sellers as an accurate guide to weight, due to the variability in how animals are fed. This pattern reveals the importance of being able to fatten animals prior to sale in order to obtain better prices.

Market price data obtained on camels and goats is not sufficient to analyse patterns, except to note that goats, being smaller than sheep, sell for less. A camel is worth between five to seven sheep in market value to producers, and was sold at an average equivalent of US$130 in spring 1999.

The price data in Table 13.2, though from a very small, un-random sample at one time period, when taken together with interview data indicates that private marketing of livestock is fairly sensitive to commercial signals. Prices are related to quality, seasonal demand and consumer preferences, indicating that marketing is more demand- than supply-driven. Distress sales do occur, typically of under-aged or breeding stock as indicated by the sale of one-year-old sheep and young female camels. But producers interviewed tried to time their annual sales to coincide with peak prices, though they are not always able to achieve this goal.

Livestock traders

Livestock trading is not an entirely new activity in Turkmenistan. Turkmen pastoralists in the Soviet period always kept some private animals which would occasionally be marketed through middlemen. The shift since independence to increasing privatisation of livestock ownership has been accompanied by the growth of the private sector in live animal trading. As the state retreats from centralised collection of livestock from producers and distribution to consumers, this niche has been filled by entrepreneurs. They provide a valuable service in giving remote producers an opportunity to realise the cash value of their livestock through the market and second, by providing meat to urban consumers.

Livestock traders operate both in rural areas and in the main market cities. There are several types of trader. Urban-based traders wait in the livestock markets for shepherds bringing animals in to sell. Since shepherds do not have time to wait around in the market for their animals to be bought, these traders buy the animals, which are often in poor condition, from the shepherds. The animals are then taken to nearby fields or barns around the town, and fattened on fodder for up to six months. The traders then resell the animals in the market at a profit.

Other types of traders go to the remoter desert areas to barter flour and household goods in return for live animals. It is easier for traders to obtain animals from the remoter areas where shepherd communities are scattered and distant from the main urban centre. Distance from urban markets makes it more difficult for individual shepherd families to organise transport for getting their own animals to market. Shepherds in the more accessible desert villages tend to sell their own animals directly to the cities, presumably because this is more profitable. Some traders tend to buy from the areas most distant to Ashgabat where producer prices are lower, to resell in Ashgabat where prices are highest and the trader can thus realise the best margin. For example, shepherds in the desert south of Dashauz, a city in northern Turkmenistan, prefer to sell to a trader who will resell their animals further away in Ashgabat, as this gives shepherds a higher price than if their animals were sold in Dashauz city.

Some rural-based traders also provide a shopping service for desert shepherds. The traders take orders from shepherd families for commodities which the trader purchases in town and brings back to the desert on the next trading trip, bartering animals in return. The main commodities which are brought back by these traders are flour, cooking oil, dress material, petrol and vehicle spare parts. Such trans-actions are based on trust, as the trader does not add a large mark-up to the urban commodities brought to the desert.

Many traders fatten the animals they buy from shepherds before re-selling. As shown in Table 13.2, fattened animals sell for premium prices. Traders based in town keep the trade animals with relatives a few kilometres from the town to avoid paying tax to municipal authorities. The animals are fed on barley, wheat chaff and stalks, cotton cake or lucerne. Some villagers with irrigated plots along the Kara Kum canal have begun to sell such types of fodder to traders. In summer months, animals are also fed on fruit and vegetable residues (e.g. melon rinds). Traders based in desert villagers hire shepherds to fatten up their trade animals on natural pasture as well as fodder.

The principal capital item required by a trader is a vehicle large enough to transport livestock. Traders use Russian-made trucks which can carry from forty to fifty sheep. In the last few years since the government has been converting state organisations into leaseholds, such large trucks have been leased out to private individuals, for example, those who worked as drivers for a state organisation. Access to such vehicles has encouraged more people to enter into such trade.

State marketing systems

The state, in the form of individual farm associations and the national livestock administration of Turkmen Mallory, is still involved in marketing animals, wool, pelts and milk. The following conclusions can be reached from the marketing patterns of associations in the study areas.

Sales of live animals by one of the largest associations which still receives some government subsidy are not planned on a financially-sound basis. Animals are sold monthly to meet administrative costs, rather than planning sales to coincide with peak seasonal prices as is done by the new private traders. Marketed animals are not fattened prior to sale, as the cost of fodder is said to be too expensive, although the association economist had not calculated the costs and returns. Larger-scale private traders and shepherds who were interviewed all fatten animals prior to sale as this is more profitable. Furthermore, male sheep are sold by the association administration at up to two years of age, though they can reach mature weight at one year. The reason younger animals are not sold is that associations must try to meet the government target of flock growth, by retaining older animals in the flock which allows higher flock numbers to be reported. Debts have arisen as some other associations have not been paid for past years' sales of pelts and breeding stock to other government organisations.

Details of one farm association's marketing in 1998/99 are as follows. The association sold 13,000 sheep or about one quarter of its average total holdings of about 50,000 sheep (all Karakul). Only 1,300 Karakul pelts were sold. All sales were through private marketing channels. No wool was sold for the preceding two years, and the association had 200 tonnes of unsold wool. The association had 2,300 camels in 1998, and used to sell camel milk, but in recent years has ceased as the milk cooling equipment no longer functions. Some 300 camels were sold. The association was owed 404 million manat (US$21,000) by several government factories and farms, to which this association sold Karakul sheep or pelts in the past two years. Karakul pelts were bought at 25,000 manat each from shepherds and sold by the association for 35,000 manat each to a private company.

In the previous year the association's administration had given its farm managers the exclusive right to buy association sheep from employed shepherds who no longer receive a salary from the association. These Karakul sheep are each valued by the association at 150,000 manat per head, but can be resold at an average price of 250,000 manat/head in the city market. This profit is in lieu of the farm managers' salary from the association.

Overall, the future for marketing livestock and animal products does not look promising for the former state farms which have become associations. Income from sales is not being used to support the remaining animals belonging to the associations. As detailed in Chapter 9, associations in the study areas are supplying few or no inputs to livestock production. However, the new associations still have considerable overheads, not least of which are the salaries of staff, such as directors, farm managers, zootechnicians, agronomists, drivers and secretaries. It appears as though much of the income derived from animals is ploughed into salaries for administrative and other staff. Some association professional staff are outspoken in their concern that association animals are being consumed as wages for workers and slaughtered on demand by elites within the higher administration. These staff foresee the time when the number of state-owned animals in the associations has been reduced to an uneconomic level, and their own jobs can no longer be financed. This scenario is similar to that which occurred in Kazakstan after privatisation in the mid-1990s (see Chapter 5).

Turkmen Mallory, the national state organisation in charge of livestock, is also involved in marketing livestock products collected from associations throughout the country. Since associations are still state organisations and their animals only leased to shepherds, the association administrations must submit some of their share of annual production to Turkmen Mallory.

A commercial centre was set up in 1998 within Turkmen Mallory in Ashgabat, with a branch office in Chargeux which deals with Karakul pelts. The function of the commercial centre is to arrange licences for foreign companies to purchase and export wool and pelts. There are at least twelve steps to be followed before an exporter can be issued with a licence, and senior staff describe the process as difficult. Exports are controlled by the Turkmen Stock Exchange, which requires the value of exports to a country to be matched by imports from that country.

Record-keeping by Turkmen Mallory's commercial centre was not well organised, but the following details were obtained. In 1998 a Swiss company had a contract to export 2,335 tonnes of wool to India via Iran, matched by importing coffee to Turkmenistan. However, the company only exported one-tenth of this quantity as the wool was unwashed and Iran refused to allow its transit through their country. As a result, the Turkmen government now requires that exported wool be washed first in the Mary wool factory. A Russian company also had a contract to export Merino sheepskins but was only able to export a quarter of this quota, then cancelled the balance of the contract. A private Turkmen company now buys most of the Karakul sheep pelts in the country, mainly through the state Karakul organisation in the northern city of Chargeux, using the Turkmen Mallory commercial centre to obtain licences. These pelts are exported to Russia for processing and manufacture into garments.

The selling prices quoted by Turkmen Mallory for different types of wool are set by the Stock Exchange, based on world prices including the cost of wool production and transport by associations. These prices are well above local market prices, as shown in Table 13.3. Producers selling directly to traders or consumers receive even lower prices. This suggests that Turkmen Mallory is adding a large mark-up to the cost of production, and that the state is undervaluing the wool harvest which association shepherds have to return to their administrations. However, the situation is more complex than this. Turkmen Mallory is unable to sell some of the association wool products at their set price; for example, 3.2 tonnes of camel wool remain unsold, and one association in the study area has 200 tonnes of unsold sheep wool. Moreover, private buyers are able to negotiate lower prices with association administrations if they have permission from the local government and approval by the Stock Exchange. Further details on wool marketing in the private sector are given below.

The market in meat is also not entirely free from state intervention, at least within Ashgabat. A government body connected with the state food supply organisation advances money to a few private traders to buy up state and privately-owned animals to sell at the Ashgabat markets. If the price of privately butchered meat rises in these markets, the traders are instructed by the government to immediately slaughter more animals and sell at a lower price. This effectively undercuts the private butchers, forcing them to reduce their prices. According to the trading agents, the government has two reasons for this; first, to ensure that

Table 13.3 State and open market prices for wool, April 1999 and May 2000

Commodity	Turkmen Mallory selling price US$ per kg	Ashbabat market selling price US$ per kg	Desert producer prices US$ per kg
Saraja unwashed white wool	0.35	0.10*	0.09 (1999) 0.27 (2000)
Saraja washed white wool	1.20	0.50	n/a
Karakul unwashed wool	0.25	0.15	0.08 (2000)
Karakul washed wool	0.50	0.30	n/a
Camel unwashed wool	3.10	0.40	0.05
Camel washed wool	n/a	0.60	n/a
Goat unwashed fibre	0.45	n/a	0.27–0.54 (2000)

Note
* This was the price paid to association administrations by a large government wool factory in Ashgabat.

meat prices do not rise too high, which could be politically sensitive, and second, the state organisations need to dispose of their livestock as quickly as possible as they need the cash income. Such trading agents are not allowed to fatten the animals before resale, in contrast to the common practice of entirely private traders, as this might lessen the supply of meat to the market. However, the agents point out that they sell some government animals to buyers for fattening to resell at a profit. Thus in addition to creating some price distortions, this policy is effectively deflating government revenue from livestock.

Private marketing of wool and pelts

In the Soviet period, wool and pelts were collected by state organisations which processed and distributed the products to the rest of the Soviet Union. Wool from privately owned animals was processed for home use as garments, mattresses and carpets, and exchanged at a much higher value than at present (see Table 13.1). As discussed above, the state marketing systems are still in place but handle a much lower volume than in the past, as the export market for wool and sheep pelts is weak. The official set prices for wool, in addition to export requirements, may also make Turkmenistan wool uncompetitive for potential exporters.

Government wool processing factories are being privatised, while at the same time several private wool and carpet-making factories have been constructed in the last couple of years and are now exporting. The export demand for Turkmen handmade carpets appears good, while the low domestic cost both of wool and skilled female labour for carpet-making means this can be a profitable business.

The export market in Karakul pelts declined sharply in the last decade, as links to Russian markets were broken and demand within Russia dropped due to domestic economic conditions. At the end of the Soviet period, Turkmenistan exported around one million Karakul pelts. Currently, a government processing factory in Chargeux still exports Karakul pelts, but a private Turkmen firm is now the main exporter of pelts from the country. Lack of technical skills and equipment to meet world standards is a constraint to further development of this private company, which has to export the raw product to Russia where further market value is added.

This private firm started in 1988 to buy Karakul pelts and export to Russia. The company has thirty-two employees and three branch offices in Turkmenistan. Pelts are bought from associations as well from private Karakul sheep farmers mainly in Mary, Lebap and Balkan regions, where most Karakul sheep are raised. The company also buys pelts from the government factory in Chargeux, selecting only the first- and second-grade pelts. In 1998 the company exported 150,000 pelts, as well as buying 70,000 from Afghanistan at a higher price to producers there than offered by the Kabul government.

The receiving company in Moscow processes the pelts, as only 15 per cent of Turkmenistan's pelts can be processed in-country, with a quality of processing not as good as in Russia. Pelts exported to Moscow are made up into fashionable garments, which is not possible in Turkmenistan as skills are lacking. The Turkmen company lacks equipment and information on how to process pelts to a higher international standard. The director is not interested in taking credit as the interest rates are too high.

The average price of Turkmenistan pelts sold at the 1998 auctions in Germany was US$18 before the August 1998 Russian financial crisis, and US$12 after the crisis. Turkmenistan pelts are mainly used to make army caps, as the highest grade fashion colour is not produced in sufficient quantity in Turkmenistan.

Marketing of wool is being privatised by the government as larger-scale private entrepreneurs are starting to invest in the wool business. A large state wool-washing and processing factory has been semi-privatised, leased since 1991 under a ten-year contract to the factory director. The factory employs between 600 and 700 people. The factory bought 3,000 tonnes of unwashed Saraja wool from farm associations in 1998 at a price of US$0.10 per kg, compared to the Turkmen Mallory official price of US$0.35 per kg. This factory also dyes and spins the wool, supplying the yarn to new private carpet-making factories in the country. The factory also supplies hand-woven carpets for sale to Turkey and Europe, exporting 6,400 m² of carpets at an average price of US$135 per m², in 1998.

The largest wool-processing plant is in Mary city. This plant was operating at about 70 per cent capacity in 2000, mainly due to non-functioning equipment dating from the 1930s. Only a small proportion of the cleaned wool is used within the country – 10 to 15 per cent, while the rest was exported. There were no plans to privatise the factory, though some foreign companies had expressed interest in investing.

Private wool-processing factories are being established to replace state plants. For example, a new factory was built in 1998 by four brothers with funds earned working in the Arabian Gulf. This factory employs 200 women to wash, dye, spin and weave wool into carpets, using twenty large looms. In 1998 this factory exported 2,000 tonnes of washed wool to Pakistan and 300 tonnes to India. By early 1999, 280 tonnes had been ordered by a Russian firm. The factory owners plan to produce carpets on order from Germany, using vegetable-dyed wool which results in more highly priced carpets. The company recently brought an expert from Pakistan to advise on using vegetable dyes. The factory buys wool from government outlets, negotiating deals below the official set price.

Private domestic carpet-making

One of the main values of sheep in the past was in the wool used to make knotted tribal carpets (*halle*) for which Turkmen tribes are internationally renowned. Trade in these carpets from the desert formerly passed through Bokhara into Russia and onto the rest of the world. Carpet-making is still carried out by some women in villages, for private sale or for family use. While nearly all women in the desert make felt mats (*ketche*) from sheep and goat wool, which are used as carpeting in the home, only a few women know how to make knotted carpets. The skills are handed down from mother to daughter in families with a reputation for carpet-making. Natural dyes are no longer used, as women consider the preparation of such dyes too labourious, though older women recognise that the natural dyes were less prone to fading than the Russian-made chemical dyes now used. Older women also note that their daughters no longer do the fine work done in former times, as the prices received for their work do not warrant the extra labour. Antique pieces are treasured as family heirlooms, though unscrupulous traders are now buying these up at a fraction of their resale value in Europe from desert families desperate for cash income.

The quality of carpet design and materials deteriorated during the Soviet period, such that modern carpets from Turkmenistan are no longer valued internationally. There is potential to re-develop traditional carpet-making skills, as has been done in Pakistan, among Turkmen refugees from Afghanistan. Currently no effort is made by government organisations in Turkmenistan to promote better-quality carpets to meet the highly discriminating tastes of the international carpet market.

Few women in the study area sell their carpets in the markets; one family sold a carpet to a government factory in 1998 at a price equivalent to US$12 per m^2 at 1998 exchange rates. She will not do so again as the factory delayed a long time in paying her. A carpet takes up to sixty days for one woman to make, though women usually work in groups. The main Ashgabat market is crowded with women trying to sell carpets. Supply seems to exceed local demand from a few foreigners and the local elite, while new Turkmen carpets are being retailed in Europe at a price of up to US$800 per m^2. The government-owned factory in Ashgabat sold carpets in 1998 to Europe at a wholesale price of US$135 m^2. The skilled Turkmen carpet-makers are not realising the potential market value of their products, as they are not able to connect with the lucrative export markets or meet international demands.

Conclusions

Marketing of livestock and their products is being increasingly privatised as a result of government policy as well as by default, as the state no longer exerts complete control over product distribution channels. The state still intervenes in price mechanisms and through financial support to state organisations involved in raising or selling livestock products. The market is not therefore totally decontrolled. Willing buyers and sellers within both government and private sectors find ways to get around the state controls. Knowledgeable informants suggest that deals can always be done and prices are negotiable, as is also implied from data presented here showing the differentials between official and actual prices paid.

State organisations in the livestock sector are gradually being phased out, but find themselves in a difficult situation common in transitional economies. They are desperately starved of operating capital while retaining obligations to pay staff and other overheads. Their marketing strategies are therefore aimed at maintaining short-term cash flow. These organisations also lack modern business practices – for example cost/benefit analysis – to run efficiently under market conditions. They equally lack processing facilities required to market their commodities profitably. They are required to value livestock products at unrealistic prices in relation to the open market, which encourages under-the-table dealing. Being state organisations, they are also required to fulfill sometimes contradictory policy targets such as increasing livestock numbers as well as the area of wheat cultivated, which means reducing fodder production as irrigated land area is limited. As state organisations, they are also prey to demands from higher authorities to supply goods without payment. Officials in state organisations are usually quite candid about all these shortcomings, but unsure of how to improve the situation. Under such conditions it is not possible for state livestock organisations to reorganise themselves into financially viable entities.

Meanwhile the private sector, both at the level of small-scale traders and large-scale manufacturing firms, is expanding. Therefore, one might expect that the private sector will soon take over the remaining market segments still controlled by state organisations. There is, however, a tendency for the government to curtail successful commercial activities through imposition of heavy taxes, burdensome export and currency regulations, and other measures. This may be an attempt to repress the rapid growth of a vigorous private sector, which is difficult to control and can compete with the state. Success in commerce necessitates close connections with state officials.

State marketing organisations are not oriented to commercial objectives, and are not pro-active in seeking new markets. They are inefficient but powerful competitors to the private sector.

For shepherds, the greatest problem in marketing their livestock products lies in the extremely low price for wool paid to producers. Wool from the Saraja Turkmen sheep breed is considered by national experts as some of the best wool for carpets. While there is demand for this type of wool from other countries (e.g. India and Pakistan), competitive commercial channels did not develop within Turkmenistan which offered producers a better price. In 2000, the export of white Saraja carpet wool was prohibited. Wool bought from producers is sold at much higher prices to other national organisations through the state marketing system. Moreover, producers now have to sell their sheep wool directly without benefit of cleaning or grading, as these functions are no longer performed by the state farms. Sale of unsorted and dirty wool further devalues the price received by producers.

Association livestock professionals and shepherds argue that creation of village-level wool-processing facilities would improve the quality and thus the price for wool received by producers. While several private wool-processing factories have opened up very recently, these are located in the cities. There is still a need for newly-privatised shepherds to be able to grade and clean their wool before selling on to commercial firms, in order to obtain higher returns. Local processing facilities could include dying and spinning wool for making into carpets, which would further raise the value of wool to shepherd families.

For entrepreneurs involved in marketing live animals, there appear to be few constraints. There is a steady domestic demand for live animals and profit margins are attractive. Trade is relatively unrestricted, and costs are mainly in buying spare parts for the ancient trucks used to transport animals. As profits increase, more successful traders will undoubtedly be able to upgrade their vehicles. However, all traders mention the difficulty of obtaining good-quality fodder to fatten animals for resale. High-quality feed concentrate is no longer available, and traders find that residues of grain and cotton processing contain fewer nutrients than previously in the Soviet period. Processing and storage of fodder crops is also less than ideal. While a private market in fodder has developed, the quality and cost of fodder production could be further improved with technical support.

Businesses involved in exporting wool and pelt (Karakul) products face a number of challenges at present. Interest rates are high while technical and marketing expertise is lacking. The owners of such businesses are keen to make international contacts which would increase their sales and profits, but do not always know how to make these connections. Provision of technical and marketing information, as well as low-interest loans, would be a vital step towards helping these new private firms to become more commercial.

In summary, there is both need and potential for developing the commercial livestock marketing sector in Turkmenistan. The government is steadily giving up control, as production and marketing have become privatised. Small-scale shepherds need to be able to gain more value from selling wool and wool products, if they are to be able to remain in production. Traders need to be able to buy better fodder to provide urban consumers with higher-quality animals. Business people need information, advice and credit to take full advantage of the new commercial environment. The deserts of Turkmenistan have long been able to support livestock whose wool and pelts are highly valued elsewhere. This capacity should not be wasted in the future.

APPENDICES

Appendix I Definition of fodder units

Definition

Fodder unit is a unit for measurement and comparison of total nutritive value.

Soviet fodder unit

Developed by Prof. E.A. Bogdanov in 1922–3, one fodder unit in this system is equal to one kg of average dry oat grain which produces 150 g of fat or 1,414 Kcal of energy in the body of an adult bull. It does not matter what kind of substances of the fodder (carbohydrates, oils or proteins) are used for fattening of animals, but composition of organic substances is important when production of milk or wool is concerned. So the precise productive equivalent of the Soviet fodder unit was developed mostly for the fattening of animals. There were also attempts to develop fodder units for the production of milk, wool, muscle work, embryo development, and growth of young animals. Further information is given in Popov (1957) and Dmitrichenko and Pshenichnyi (1964).

Fodder unit is used for weight units of fodder (kg, quintal, ton) or productivity of one hectare and expressed in grams, kilograms, quintals or tons.

Appendix II Detailed sampling methodology

Production of pasture fodder (biomass) is expressed in air-dried mass or absolutely dry mass. Samples in our experiments for assessing productivity were obtained by cutting the plants, distinguished from the zootechnical method when consumption of vegetation by animals is measured.

Plants are cut at a level of five cm above ground without selection of 'typical' vegetation stands. Areas of record plots are not less than ten square metres. Samples can be taken from one place (10 m^2) or different places (1 m^2 to 10 m^2 and 2.5 m^2 to 4 m^2 and more).

After cutting, the vegetation mass is weighed on site. An average sample of one kg is taken from the whole vegetation mass and used for the determination of dry mass (dry matter), botanical composition, and chemical composition. The sample is then dried to a state of permanent air-dried matter, assessed by repeated weighing. Then the amount of dry mass is determined followed by calculation of the percentage of dry mass (e.g. 11 to 13 per cent is the content of air-dried mass).

Botanical composition

Each average sample of one kg described above is used for determining the botanical composition. The sample is divided into fractions composed of dominant plants, associated plants, weed plants and plants which are unpalatable and poisonous at the season of sampling. Then determination of the percentage of each fraction is followed by calculation of productivity of each fraction multiplied up to one hectare.

A transect method is used for areas such as sands with very sparse vegetation. The size of a transect on the semi-shrub pastures with ephemerous vegetation such as the transect in the Taukum sand desert is $300\,m^2$ (2 m to 150 m). Preferably the transect should be mapped (to map plant aggregations). Each plant aggregation has its own biomass productivity. Therefore productivity is measured at each plant aggregation's record plot of $10\,m^2$. Sizable shrubs along a transect are counted and measured (height, diameter, one-year shoots). Biomass is estimated from the ground surface to the height of 120 cm and only one-year shoots are included in the estimation.

The following investigations are carried out at a transect:

- determination of the elements of a plant complex and their ratio;
- calculation of sizable shrubs;
- mapping of different plant aggregations;
- estimation of biomass;
- determination of productivity of one ha of pasture.

Preparation of samples for analysis

Each sample of one kg used for determining dry matter and botanical composition is also used for the determination of the chemical (zootechnical) analysis. The sample is ground and sifted through a sieve with pore size of 0.5 mm. The resulting air-dried substance is used for selection of a laboratory sample. This substance is spread on a surface in the form of a square and divided by horizontal and vertical lines into small squares. Samples for chemical analyses are taken from the resulting small squares. The weight of the laboratory sample depends on the amount of analyses to be done and ranges from 50 g to 200 g.

These methods are based on the following: Dospekhov 1968; *Experimental Methods* 1971; *Field Geobotany* 1959, 1960 and 1964; Lukoshyk 1965; Nechaeva 1957; Neidin 1968; Zhuravlyov 1963.

BIBLIOGRAPHY

ADB (Asian Development Bank) (1996a) Strengthening the Implementation of Agriculture Sector Reforms. *Working Paper No. 1 on Land Reform.* TA No. 2356. Almaty: Danagro Adviser A/S and Landell Mills Ltd.

ADB (Asian Development Bank) (1996b) Legal and institutional reform in the agricultural sector. *Study of Market Reforms in the Agricultural Sector* (R. Gaynor). TA No. 7. Almaty Danagro Adviser A/S and Landell Mills Ltd.

ADB (Asian Development Bank) (1997) *Wool Marketing and Production: Transition from Set-Back to New Growth* (R. Diddi and M. Menegay). Washington, DC: Abt Associates.

Akiner, S. (1995) *The Formation of Kazatch Identity: From Tribe to Nation-State.* London: The Royal Institute of International Affairs.

Alimaev, I.I. (1986) *Conditions of Pastures in Almaty and Measures for Raising and Preserving their Productivity.* Report to the Oblast Committee of the Communist Party. Almaty [in Russian].

Alimaev, I.I., Zhambakin, A. and Pryanishnikov, S.N. (1986) Rangeland farming in Kazakstan. *Problems of Desert Development* **3**: 14–19.

Aryngaziev, S. (1998) Downy goats in Kazakstan. *European Fine Fibre Network* **3** (June): 13–15, Aberdeen: Macaulay Land Use Research Institute.

Asanov, K.A., and Alimaev, I.I. (1990) New forms of organisation and management of arid pastures of Kazakstan. *Problems of Desert Development* **5**: 42–9.

Asanov, K., Alimaev, I.I. and Smailov, K.S. (1992a) Effect of grazing on soil and plant covers in Northern Kazakhstan Desert. *Problems of Desert Development* **2**: 7–13.

Asanov, K., Shax, B.P., Alimaev, I.I. and Pryanishnikov, S.N. (1992b) *Pasture Sector of Kazakstan.* Almaty: Ghylym [in Russian].

Ataev, M. (1999) Reforms in the agrarian sector of Turkmenistan's economy. *Russian and East European Finance and Trade* **35**(5): 75–95.

Babaev, A.G. (1994) Landscape of Turkmenistan. In V. Fet (ed.) *Biogeography and Ecology of Turkmenistan.* Dordrecht: Kluwer Academic Publishers, pp. 5–21.

Babaev, A.G. and Kharin, N. (1992) Map of desertification of arid territories of Asia. *Problems of Desert Development* **5**: 37–39.

Babaev, A.G. and Orlovsky, N.S. (1985) Natural conditions of deserts in the USSR. In N.T. Nechaeva (ed.) *Improvement of Desert Ranges in Soviet Central Asia.* Chur, Switzerland: Harwood.

Babaev, A.G., Nikolaev, V.N. and Orlovski, N.S. (1991) Current status and prospects for use of natural forage lands and dry farming in the Aral Sea basin. *Problems of Desert Development* **6**: 3–11.

Banks, T. (1999) State, community and common property in Xinjiang: synergy or strife. *Development Policy Review* **17**: 293–313.

Behnke, R.H. (1983) Production rationales: the commercialization of subsistence pastoralism. *Nomadic Peoples* **14**: 3–33.

Behnke, R.H. (1999) Pers. comm. Pastoral development specialist, Macaulay Land Use Research Institute, Aberdeen.

Behnke, R.H. and Scoones, I. (1993) Rethinking range ecology: implications for rangeland management in Africa. In R.H. Behnke, I. Scoones, and C. Kerven (eds) *Range Ecology at Disequilibrium: New Models of Natural Variability and Pastoral Adaptation in African Savannas.* London: Overseas Development Institute

Behnke, R.H., Scoones, I., and Kerven, C. (eds) (1993) *Range Ecology at Disequilibrium: New Models of Natural Variability and Pastoral Adaptation in African Savannas.* London: Overseas Development Institute.

Beloborodova, G.G. (1964) Meteorologicheskie usloviya i yrozhai pastbish'noi rastitel'nosti Betpak-daly [Meterological conditions and yeilds of rangeland plants of Betpak-dala]. *Trudy Instituta Botaniki ANKazSSR [Proceedings of the Botanical Institute of the Kazakhstan Acadmy of Sciences]* **18**: 87–112.

Benson, L. and Svanberg, I. (1998) *China's Last Nomads: The History and Culture of China's Kazaks.* New York: M.E. Sharpe.

Bloch, P. and Rasmussen, K. (1997) Land reform in Krygyzstan. In S. Wegren (ed.) *Land Reform in Central Asia.* London: Routledge.

Brown, C.G. and Longworth, J.W. (1995) Wool: China, change and trade. *Current Affairs Bulletin* (Sydney, Australia) (Aug/Sept): 4–13.

Bruzon, V. (1998) Range Management Report. For pilot project on improvement of livestock (cattle and sheep). Ashgabat: TACIS/SATEC International.

Calef, W. (1960) *Private Grazing and Public Lands: Local Management of the Taylor Grazing Act.* Chicago: University of Chicago Press.

Carana Corporation (1996) *Kazak Meat and Dairy Systems Analysis.* First follow-up report for USAID, Almaty.

Caughley, G., N. Sheperd and J. Short (eds) (1987) *Kangaroos: Their Ecology and Management in the Sheep Rangelands of Australia.* New York: Cambridge University Press.

Central Asian Review (1962) Private property tendencies in Central Asia and Kazakhstan. **10**(2): 147–56.

Channon, J. and Channon, S. (1990) *The livestock sector in the USSR with particular reference to Soviet Central Asia: A bibliographical study of works in Russian from 1917 to the present day.* Manuscript for Institute of Development Studies, Brighton, University of Sussex.

Collins, P. and Nixson, F. (1993) Managing the implementation of 'shock therapy' in a land-locked state: Mongolia's transition from the centrally planned economy. *Public Administration and Development* **13**: 389–407.

Coppock, D.L. (1993) Vegetation and pastoral dynamics in southern Ethiopian rangelands: implications for theory and management. In R.H. Behnke, I. Scoones and C. Kerven (eds) *Range Ecology at Disequilibrium.* London: Overseas Development Institute.

Dhambul District Administration, Land Use Department, Almaty Oblast (1991) An analysis of production and financial activity of kolkhozes and sovkhozes in Dhambul District for 1986–1900. District Administration, Uzunagach [in Russian].

—— (1992) An analysis of production and financial activity of kolkhozes and sovkhozes in Dhambul District for 1991. Raion Administration, Uzunagach [in Russian].

—— (1998) Report on animal husbandry for December, form No. 24. Uzunagach, Dhambul Raion, Almaty Oblast [in Russian].

—— (1998) Lands held by farms in Dhambul Raion as of 1 January 1998. Uzunagach, Dhambul Raion, Almaty Oblast [in Russian].

Dhambul District Administration, Almaty Oblast (1998) Private livestock in Dhambul Raion as of 1 February 1998. Mimeo. Raion Administration, Uzunagach, Dhambul Raion, Almaty Oblast [in Russian].

Dhambul District Administration, Agricultural Statistics Services (n.d.) Unpublished records on Kurtinsky Raion and Dhambul Raion livestock and arable production [in Russian].

Dmitrichenko, A.P. and Pshenichnyi, P.D. (1964) *Feeding of Agricultural Animals*. Leningrad: Nauka [in Russian].

Dospekhov, B.A. (1968) *Field Work Methods*. Moscow: Kolos [in Russian].

Ellis, J.E. (1970) A computer analysis of fawn survival in the Pronghorn. Ph.D. dissertation. Dept. of Zoology, University of California, Davis.

Ellis, J.E. (1994) Climate variability and complex ecosystem dynamics: implications for pastoral development. In I. Scoones (ed.) *Living with Uncertainty*. London: Intermediate Technology Publications.

Ellis, J. and Galvin, K. (1994) Climate patterns and landuse practices in the dry zones of Africa. *Bioscience* **44**(5): 340–9.

Ellis, J.E. and Swift, D.M. (1988) Stability of African pastoral ecosystems: alternate paradigms and implications for development. *Journal of Range Management* **41**(6): 450–9.

Ellis, J.E., Galvin, K., McCabe, J.T. and Swift, D.M. (1987) *Pastoralism and Drought in Turkana District, Kenya*. Report to the Norwegian Aid Agency for International Development, Nairobi.

Ermekov, M.A. and Golodnov, A.B. (1976) Degeres sheep. In Mynbaevo Research Institute of Sheep Breeding, *Fat Tail Sheep of Kazakstan*. Alma Ata: Kainar [in Russian].

Experimental Methods at the Hayfields and Pastures, vols I and II (1971). Moscow: Kolos [in Russian].

Fadeev, V.A. and Sludskii, A.A. (1982) *Saiga v Kazakstane [The Saiga in Kazakstan]*. Alma Ata: Nauka [in Russian].

Falkingham, J. (1999) Poverty in Central Asia. In *Central Asia 2010 Prospects for Human Development*. New York: UNDP.

Falkingham, J., Klugman, J., Marnie, S. and Micklewright, J. (eds) (1997) *Household Welfare in Central Asia*. London: Macmillan Press.

FAO (Food and Agriculture Organisation) (2000) *FAOSTAT Database*. Web site: http://apps.fao.org

Fernandez-Gimenez, M. (1999) Reconsidering the role of absentee herd owners: a view from Mongolia. *Human Ecology* **27**(1): 1–27.

Field Geobotany, vol. I (1959). Moscow, Leningrad: Nauka [in Russian].

Field Geobotany, vol. II (1960). Moscow, Leningrad: Nauka [in Russian].

Field Geobotany, vol. III (1964). Moscow, Leningrad: Nauka [in Russian].

Finke, P. (1995) Kazak pastoralists in western Mongolia: economic and social change in the course of privatisation. *Nomadic Peoples* **36/37**: 195–216.

Foss, P.O. (1960) *Politics and Grass*. Seattle: University of Washington Press.

Gilmanov, T. (1995). The state of rangeland resources in the newly-independent states of the former USSR. In N.E. West (ed.) *Rangelands in a Sustainable Biosphere. Proceedings of the*

Fifth International Rangeland Congress, Vol. II, Denver: Society for Range Management, pp. 10–13.

Goskomstat [State Statistical Committee] (1984–8) *Narodnoe Khozyastvo Kazakstana, Statisticheskii Yezhegodnik [The Economy of Kazakstan, Statistics Yearbook]*, Alma Ata [in Russian].

Goskomstat [State Statistical Committee] (1991) *Annual Regional Statistics of Kazakstan.* Almaty: Republic Information Publishing Centre [in Russian].

Goskomstat [State Statistical Committee] (1996, 1998) *Selskoe Khozyastvo Respublika Kazakstan [Agricultural Statistics of the Republic of Kazakstan]*. Almaty [in Russian].

Government of Kazakstan (1995) Presidential Decree Number 2235 Having the Force of Law: On Taxes and Other Mandatory Payments to Revenue. 24 April 1995.

Government of Kazakstan, State Committee on Statistics and Analysis (1997/8) *Statistical Bulletins*. Almaty [in Russian].

Government of Mongolia, Ministry of Agriculture and Industry (2000) Web site: www.pmis.gov.mn/agriculture

Goyal, H.D. (1999) A development perspective on Mongolia. *Asian Survey* **39**(4): 633–55.

Gray, K. (ed.) (1990) *Soviet Agriculture: Comparative Perspectives*. Ames: Iowa State University Press.

Green, D.J. and Vokes, R. (1997) Agriculture and the transition to the market in Asia. *Journal of Comparative Economics* **25**: 256–80.

Griffin, K. (1999) The role of the state in transition. In *Central Asia 2010: Prospects for Human Development*. New York: UNDP, pp. 40–7.

Haghayeghi, M. (1996) Kazakstan's declining agriculture. *Central Asian Monitor* **1**: 15–18.

Haghayeghi, M. (1997) Politics of privatisation in Kazakstan. *Central Asian Survey* **16**: 321–38.

Hamann, B. (1999) Kazakhs in Xinjiang, China. Unpublished data from Ph.D. thesis 'Ecological and socioeconomic impacts of the Chinese reform policy within the Southern border zone of the Dsungarian basin, Xinjiang, China'. Landscape Planning Department, Technical University of Berlin.

Hesse, C. and Trench, P. (2000) *Who's Managing the Commons?* London: IIED.

Howell, J. (1995) Household coping strategies in Kyrgyzstan. *Development in Practice* **5**(4): 361–4 (Oxford: Oxfam).

Institute of Ethnography (1973) *History of the Economy of the Peoples of Central Asia and Kazatchstan* (TIE vol. XCVIII). Leningrad: Nauka [in Russian], pp. 207–22 .

International Monetary Fund (IMF) (1999) *Economic Reforms in Kazakstan, Kyrgyz Republic, Tajikistan, Turkmenistan and Uzbekistan*. Occasional Paper 183. Washington, DC: IMF.

Joint BIS–IMF–OECD World Bank (2000) Statistics on external debt. http://www.oecd.org/dac/debt

Justice, C., Hall, D., Salomonson, V. *et al.* (1998) The Moderate Resolution Imaging Spectroradiometer (MODIS): land remote sensing for global change research. *IEEE Transactions on Geoscience and Remote Sensing* **36**(4): 1228–49.

Kalyuzhnova, Y. (1998) *The Kazakstani Economy: Independence and Transition*. Reading: University of Reading.

Kanapin, K. and Jumadillaev, K. (1983) Degeres sheep of Mynbaeva Research Station and inheritance of productive traits in crossbreeds with Edilbaev sheep, In Mynbaevo Research Institute of Sheep Breeding, *Increasing of Wool and Meat Production of Sheep*. Alma Ata [in Russian] pp. 9–100.

Karaganda Zemliustroistvo [Karaganda land office] (1998) *Data on Land Allocation for Private Farms and Collectives*. Karaganda: Karaganda Land Office [in Russian].

Kazgiprozem [Kazakhstan Institute for Land Planning] (1981). *Land Use and Organisation of Sovkhoz Zhenis, Zhana-arkin raion, Dzhezkazgan oblast*. Dzhezkazgan: Ministry of Agriculture [in Russian].

Kazgiprozem (1988) *Natural Forage Resources of Sovkhoz Sarysu, Ulutau raion, Dzhezkazgan oblast*. Dzhezkazgan: Ministry of Agriculture [in Russian].

Kerven, C. (1992) *Customary Commerce: A Historical Reassessment of Pastoral Livestock Marketing in Africa*. ODI Occasional Paper 15. London: Overseas Development Institute.

Kerven, C. and Alimaev, I.I. (1998) Mobility and the market: economic and environmental impacts of privatisation on pastoralists in Kazakstan. Paper presented at the conference on 'Strategic Considerations for the Development of Central Asia', Council for Development of Central Asia and Chinese Academy of Science. Urumchi, China, September 13–18.

Kerven, C., Channon J. and Behnke, R. (1996) *Planning and Policies on Extensive Livestock Development in Central Asia*. Working Paper 91. London: Overseas Development Institute.

Kerven, C., Lunch, C., and Wright, I. (1998) Impacts of privatisation on livestock and rangeland management in semi-arid Central Asia, *Report on Field Trip to Kazakstan and Turkmenistan, Jan 21–March 10 1998 for the Overseas Development Institute*. London: Overseas Development Institute.

Kerven, C., Russel, A. and Laker, J. (2000) The potential for increasing producers' income from wool, fibres and pelts. Unpublished report for Macaulay Land Use Research Institute and International Livestock Research Institute, Aberdeen/Nairobi.

Khalova, G.O. and Orazov, M.V. (1999) Economic reforms and macro-economic policy in Turkmenistan. *Russia and East European Finance and Trade* **35**(4): 7–44.

Khazanov, A. (1984) *Nomads and the Outside World*. Cambridge: Cambridge University Press.

Kirichenko, H.G. (1980) *Pastbisha pustyn Kazakstana* [*Desert pastures of Kazakstan*]. Alma Ata: Nauka [in Russian].

Kirichenko, N.G. (1966) Dinamika urozhaya i khimizma osnovykh pastbish pustyni Betpak-daly [Yield dynamics and chemical content of pastures of Betpak-dala]. *Trudy Instituta Botaniki AN KazSSR* [*Proceedings of the Botanical Institute of the Kazakhstan Acadamy of Sciences*] **23**: 3–53 [in Russian].

Le Houerou, H.N. (1984) Rain-use efficiency: a unifying concept in arid-land ecology. *Journal of Arid Lands* **7**: 213–47.

Libecap, G.D. (1981) *Locking up the Range: Federal Land Contrals and Grazing*. Cambridge, MA: Ballinger Publishing Company

Longworth, J.W. and Williams, G.J. (1993) *China's Pastoral Region: Sheep and Wool, Minority Nationalities, Rangeland Degradation and Sustainable Development*. Wallingford: CAB International.

Lukoshyk, N.A. (1965) *Zootechnical Analysis of Fodder*. Moscow: Selkhozgiz [in Russian].

Mamedov, B. (1998) Pers. comm. Hydrological specialist. National Institute of Deserts, Flora and Fauna (NIDFF). Ashgabat.

Matley, I.M. (1994) The population and the land, in E. Allworth (ed.) *Central Asia: 120 Years of Russian Rule*, Durham: Duke University Press, pp. 92–130.

McCauley, M. (1994) Agriculture in Central Asia and Kazakhstan in the 1980s. In S. Akiner (ed.) *Political and Economic Trends in Central Asia*. London: British Academic Press.

McGregor, B.A. (1996) *Production and Processing of Cashmere in China*. Report on a study tour. Victoria, Australia: Victoria Institute of Animal Science.

Meat and Livestock Commission (1983) *Feeding the Ewe*. Milton Keynes, UK: Meat and Livestock Commission.

Mehrotra, S. (1999) Public spending priorities and the poor in Central Asia. In *Central Asia 2010: Prospects for Human Development*. New York: UNDP, pp. 48–57.

Mission of Mongolia to the United Nations (2000) Web site: www.un.int/mongolia

Nechaeva, N.T. (1957) *Methods for the Assessment of Fodder Reserve of Desert Pastures*. Ashgabat: Ilim [in Russian].

—— (1985) *Improvement of Desert Ranges in Soviet Central Asia*. Chur, Switzerland: Harwood.

Neidin, N.G. (1968) *Field Experiment*. Moscow: Kolos [in Russian].

Neupert, R. (1999) Population, nomadic pastoralism and the environment in the Mongolian Plateau. *Population and Environment* **20**(5): 413–41.

Nicholls, N. and Wong, K.K. (1990) Dependence of rainfall variability on mean rainfall, latitude, and the southern oscillation. *Journal of Climate* **3**: 163–70.

Nikolayev, V. (1980) *Chemical Composition of Pasture Forage of the Kara Kum*. Ashgabat [in Russian].

Nordblom, T.L., Shomo, F. and Gintzburger, G. (1996) *Food and Feed Prospects for Resources in Central Asia. Central Asian Livestock Regional Assessment*. Workshop Proceedings, Tashkent. Small Ruminant Collaborative Research Support Progam, Davis, California.

OECD (Organisation for Economic Co-operation and Development) (1995) *Agricultural Policies, Markets and Trade in the Central and Eastern European Countries, Selected New Independent States, Mongolia and China*. Paris: OECD.

O'Hara, S.L. (1997) Agriculture and land reform in Turkmenistan since independence. *Post-Soviet Geography and Economics* **38**(7): 430–44.

Olcott, M. (1981) The settlement of the Kazakh nomads. *Nomadic Peoples* **8**: 12–23.

Olcott, M. (1995) *The Kazakhs*, second edition. Stanford, CA: Hoover Institution Press.

Ospanova, B.C. (ed.) (1996) *Fodder Plants of the Pastures and Hayfields of Kazkahstan*. Almaty: Kainar.

Pastor, G. and van Rooden, R. (2000) *Turkmenistan: The Burden of Current Agricultural Practices*. IMF Working Paper WP/00/98. Washington, DC.

Pomfret, R. (1995) *The Economies of Central Asia*. Princeton, NJ: Princeton University Press.

Pomfret, R. (1997) Growth and transition: why has China's performance been so different? *Journal of Comparative Economics* **25**(3): 422–40.

Pomfret, R. (1998) The transition to a market economy, poverty and sustainable development in Central Asia. Paper presented at the conference on 'Strategic Considerations for the Development of Central Asia.' Council for Development of Central Asia and Chinese Academy of Sciences, Urumchi, China, September 13–18.

Pomfret, R. (1999) Development strategies and prospects for the future. In *Central Asia 2010: Prospects for Human Development*. New York: UNDP.

Pomfret, R. (2000a) Agrarian reform in Uzbekistan: why has the Chinese model failed to deliver? *Economic Development and Cultural Change* **48**(2): 269–84.

Pomfret, R. (2000b) Transition and democracy in China. *Europe-Asia Studies* **52**(1): 149–60.

Popov, I.S. (1957) *Feeding of Agricultural Animals*, 9th edition. Moscow: Kolos [in Russian].

Prince, S.D., de Colstoun, E.B. and Kravitz, L.L. (1998) Evidence from rain-use efficiencies does not indicate extensive Sahelian desertification. *Global Change Biology* **4**: 359–74.

Republic of Kazakstan Civil Code (1995) *Edict on Land Law*. Almaty [in Russian].

Rhind, S.M. (1995) Management sheep for successful breeding. *Feed Mix* **3**: 4–46.

Röhm, T. (1995) *Economic Reforms and Transition to Market Economy in Kazakhstan and the Kyrgyz Republic.* Progress report, Joint research project, Tokyo: Institute of Developing Economies and Munich: Ifo Institute for Economic Research.

Ropelewski, C.F. and Halpert, M.S. (1987) Global and regional scale precipitation patterns associated with the El Nino/Southern Oscillation. *Monthly Weather Review* **115**: 1606–26.

Running, S.W. (1990) Estimating terrestrial primary productivity by combining remote sensing and ecosystem simulation. In R.J. Hobbs and H.A. Mooney (eds) *Remote Sensing and Biosphere Functioning.* New York: Springer-Verlag.

Russel, A.J.F., Doney, J.M. and Gunn, R.G. (1969) Subjective assessment of body fat in live sheep. *Journal of Agricultural Science* **72**: 450–4 (Cambridge).

Ryder, M.L. (1983) *Sheep and Man.* London: Duckworth.

Sandford, S. (1983) *Management of Pastoral Development in the Third World.* New York: Wiley.

Sadykulov, T.S. (1985) *Degeres Sheep.* Alma Ata: Kainar [in Russian].

Sabeko, L.V. (1998). *Agriculture in Almaty Oblast 1985–1997: A Statistics Handbook.* State Committee on Statistics and Analysis, Almaty Oblast: Statistical Service [in Russian].

Sludskii, A.A. (1963) Dzhuts in Eurasian steppes and deserts. *Trudii Instituta Zoologii AN KazSSR [Proceedings of the Zoological Institute of the Kazakstan Acadamy of Sciences]* **20**: 5–88 [in Russian].

Sneath, D. (1999). Spatial mobility and inner Asian pastoralism. In C. Humphrey and D. Sneath (eds) *The End of Nomadism? Society, State and the Environment in Inner Asia.* Durham, NC: Duke University Press.

Sneath, D. (2000). Producer groups and the decollectivisation of the Mongolian pastoral economy. In J. Heyer, F. Stewart and R. Thorp (eds) *Group Behaviour and Development.* Oxford: Oxford University Press.

Spoor, M. (1996) *Upheaval Along the Silk Route: The Dynamics of Economic Transition in Central Asia.* Working Paper Series, No. 216. The Hague: Institute of Social Studies.

Spoor, M. (1997a) Agrarian transition in the former Soviet Union Central Asia: the case of Central Asia. In M. Spoor (ed.) *The 'Market Panacea': Agrarian Transformation in LDCs and former Socialist Economies.* London: Intermediate Technology Publications, pp. 29–42.

Spoor, M. (1997b) Introduction. In M. Spoor (ed.) *The 'Market Panacea': Agrarian Transformation in LDCs and former Socialist Economies.* London: Intermediate Technology Publications, pp. 29–42.

State Committee on Statistics and Analysis (1998) *Statistical Bulletin.* Almaty.

Technichal Assistance to the Commonwealth of Independent States, European Union (TACIS) (1996) Support to Farmers' Associations including Rural Credit. *Technical Assistance to the Economic Reform in Turkmenistan.* Terms of Reference. FDTUR 9601, Ashgabat.

Technichal Assistance to the Commonwealth of Independent States, European Union (TACIS) (1997) Meat production statistics for Turkmenistan, Ashgabat: TACIS.

Technichal Assistance to the Commonwealth of Independent States, European Union (TACIS) (1998) Weekly bulletin of livestock prices (Dec. 1997–Aug. 1998). TACIS Agricultural Information on Markets project. Almaty: TACIS.

Templar, G., Swift, J. and Payne, P. (1993) The changing significance of risk in the Mongolian pastoral economy. *Nomadic Peoples* **33**: 105–22.

Tiffen, M., Mortimore, M. and Gichuki, F. (1995) *Population Growth and Environmental Recovery: Policy Lessons from Kenya.* London: International Institute for Environment and Development.

Toulmin, C. and Quan, J. (eds) (2000) *Evolving Land Rights, Policy and Tenure in Africa*. London: International Institute for Environment and Development.

Trewartha, G.T. (1957) *Elements of Physical Geography*. New York: McGraw-Hill.

Tucker, C.J. and Sellers, P.J. (1986) Satellite remote sensing of primary production. *International Journal of Remote Sensing* **7**: 1395–416

ULG Consultants Ltd (1997) Agriculture in transition (A. Jones). Paper for the Kyrgyz Agriculture Sector Conference, Bishkek, December 1997.

UNDP (United Nations Development Programme) (1999) *Transition 1999: Human Development Report for Central and Eastern Europe and the CIS*. New York: UNDP.

van Veen, S. (1995) Kyrgyz sheep herders at a crossroads. *Pastoral Development Network* **28d**. London: Overseas Development Institute.

Vidon, H. (1999) *Influence de la privatisation de l'élevage sur la santé animale au Kazakstan: Le cas de l'élevage ovin*. Diplôme d'études supérieures specialisées productions animales en régions chaudes. Maisons-Alfort, France: Ecole Nationale Véterinaire d'Alfort.

Werner, C.A. (1994) A preliminary assessment of attitudes towards the privatization of agriculture in contemporary Kazakhstan. *Cental Asian Survey* **13**(2): 295–303.

Werner, C.A. (1998) Household networks and the security of mutual indebtedness in rural Kazakstan. *Central Asian Survey* **17**: 597–612.

Westoby, M., Walker, B.H. and Noy-Meir, I. (1989) Opportunistic management for rangelands not at equilibrium. *Journal of Range Management* **42**: 266–74.

Williams, D.M. (1997) The desert discourse of modern China. *Modern China* **23**(3): 328–355.

Wily, L.A. (2000) Land tenure reform and the balance of power in eastern and southern Africa. *Natural Resource Perspectives* **58**. London: Overseas Development Institute.

Wong, L.-F. and Ruttan, V. (1990) A comparative analysis of agricultural productivity trends in centrally planned economies. In K.R.Gray (ed.) *Soviet Agriculture: Comparative Perspectives*. Ames: Iowa University Press.

World Bank (1993) *Kazakhstan: Transition to a Market Economy*. Country Report. Washington, DC: World Bank.

World Bank (1997) *Kyrgyz Republic Agricultural Sector Review*. Rural Development and Environment Sector, Europe and Central Asia Region. Washington, DC: World Bank.

World Bank (2000) Europe and Central Asia Country Briefs and Development Data: Kazakstan, Kyrgyzstan, Turkmenistan and Uzbekistan. Web site: http://www.worldbank.org

Wright, I.A. (1998) *Report on Visit to Turkmenistan, August 1998*, for project 'The Impacts of privatisation on range and livestock management in semi-arid Central Asia'. London: Overseas Development Institute.

Ye, E.Q. (1999) The second stage of agriculture sector reforms in the Central Asian Republics. In *Central Asia 2010: Prospects for Human Development*. New York: UNDP.

Yu, F., Price, K.P., Lee, R. and Ellis, J. (1999) Use of time series of AVHRR NDVI composite images to monitor grassland dynamics in Inner Mongolia, China. *Proceedings of ASPRS '99* (accepted).

Zanca, R. (1999) Kolkhozes into Shirkats: A local label for managed pastoralism in Uzbekistan. Paper presented at American Ethnological Society Meetings, Portland Oregon, March 25–8.

Zhambakin, Z.A. (1995) *Pastbisha Kazakhstana [Pastures of Kazakhstan]*. Almaty: Kainar [in Russian].

Zhuravlyov, E.I. (1963) *Guide for the Zootechnical Analysis of Fodder*. Moscow: Selkhoizdat [in Russian].

Zveriakov, I.A. (1932) *From Nomadic Life – to Socialism. Proceedings and Materials, Vol.1. General Problems of Socialist Reconstruction in Agriculture.* Kazak Research Institute of Socialist Reconstruction in Agriculture. All-Union Academy of Agricultural Sciences. Alma Ata: Kraevoe Izdatelstvo Ogiza v Kazakstane – SSKVA [in Russian].

INDEX

absentee owners 182–83
administrative 90, 95–6, 188, 202; centre 85, 100, 121, 148, 168, 172, 174, 178, 221; costs 249; districts 29, 84, 172, 214; headquarters 243; levels 12, 77; records 75; region 29; structures 7
Africa: pastoralists 105–6; Sahel 56; semi-arid 96, 105
agrarian reform 10–11; comparison between Kazakstan and Turkmenistan 20; in Kyrgyzstan 13; in Mongolia 24–5; in southeast Asia and China 22; tenure 94; transformation 18; transition in Central Asia 165; in Turkmenistan 14; in Uzbekistan 15.
agricultural 9; and crop 56; economy 80, 168; equipment 28, 34, 82, 156; inputs 11; output 11, 13, 15, 16, 18, 22, 28, 105, 168; policy 210–11; production 9, 22, 80–3, 87–8, 97, 106; products 11, 26; reform 10, 187; sector 16, 20, 22–3, 107
Agropyron 28, 33, 34, 38, 41, 42, 47, 109, 113, 117, 124, 131, 141, 142, 161
Ala Tau mountains 29, 35, 36, 53, 61, 77, 133, 136
alfalfa 28, 34, 190, 215, 217, 220–1, 225
Almaty 1, 27, 29, 34, 35, 61; city 27, 85, 148, 155, 161–2; market 114, 154–5, 161; *oblast* 1, 29, 34–5, 61, 77–9, 83, 109, 119, 138, 139, 140, 148, 150–1, 155, 161–2
American ranchers 96, 106
animals: accumulate 2, 157, 164, 239, 228, 229, 239, 242; fibres 26, 27, 122, 147, 153, 163, 164, 214, 227, 251; health 108, 119, 122, 151, 186, 192, 207, 218, 232, 239, 240; products 15, 119, 149, 157, 159–60, 162–4, 230,

242–4, 250; value per head 234; *see also* livestock
Artemisia 34–7, 42, 50, 53
Ashgabat 168, 172, 177, 183, 194, 195, 215, 219, 226, 250; market 243, 246–8, 251, 254
Asian 2, 104, 106; Development Bank 16
assets: association 9, 171, 211; cooperative 83, 91, 93; farm 4, 7, 12, 82, 84, 136, 156; farmers and shepherds 125, 138, 145, 201; livestock 7, 12, 18, 87, 148; pastoral 82; productive 75, 80–1, 83, 87
associations (farmers') 7–9, 163, 168, 171–94, 200–2; administration 9; camels 224; centres 177, 178, 221, 236; costs 7; land 168, 184, 201, 215; leasing 174, 211, 223, 227, 236; livestock 7, 168, 171–3, 178, 179–80, 182, 191, 212–13, 215–16, 220, 229; management 9, 174–6, 178, 179, 180, 202; marketing 163, 176, 243, 249–51, 255; members 8, 171; pastures 8, 191, 223; produce 7, 211; resources 174; waterpoints 8, 184–5, 188–9, 202; *see also* peasant farms
autumn 3; livestock condition 116, 124; grazing and forage 113, 134–5, 203; livestock movement 28, 32, 45, 102, 109, 128, 131, 208; mating 37, 115; pastures 32, 37, 43, 45, 57, 72, 73, 136, 190, 191, 198, 216–17; plant samples 35; precipitation 3, 36, 57; prices 154; vegetation 36

barley 28, 33, 36, 113, 117, 161, 169, 177, 197, 220, 221, 226, 249
barn and barns 19, 28, 44, 85, 94, 100, 145, 160; private 90, 93, 114, 138, 157,

T - #0015 - 071024 - C0 - 234/156/16 [18] - CB - 9780700716999 - Gloss Lamination